BASIC ALGEBRA
AND GEOMETRY
MADE A BIT EASIER

BASIC ALGEBRA
AND GEOMETRY
MADE A BIT EASIER

Concepts Explained in Plain English,
Practice Exercises, Self-Tests, and Review

LARRY ZAFRAN

Self-Published by Author

BASIC ALGEBRA AND GEOMETRY MADE A BIT EASIER:
Concepts Explained in Plain English, Practice Exercises,
Self-Tests, and Review

Book design by Larry Zafran
Graphics created by author and/or are
public domain from Wikimedia Commons

Printed in the United States of America
First Edition printing March 2010
First Edition revised March 2011

ISBN-10: 1-4499-5853-2
ISBN-13: 978-1-44-995853-4

Please visit the companion website below for additional information, to ask questions about the material, to leave feedback, or to contact the author for any purpose.

www.MathWithLarry.com

CONTENTS

CHAPTER ZERO

INTRODUCTION

ABOUT THE *MATH MADE A BIT EASIER* SERIES

This is the fourth book in the self-published *Math Made a Bit Easier* series which will be comprised of at least eight books. The goal of the series is to explain math concepts "in plain English" as noted in the book's subtitle. The series also attempts to explain the truth about why students struggle with math, and what can be done to remedy the situation. To write with such candidness is only really possible as a completely independent author.

Unlike many commercial math books, this series does not imply that learning math is fast, fun, or easy. It requires time and effort on the part of the student. It also requires that the student be able to remain humble as s/he uncovers and fills in all of his/her math gaps.

THE PURPOSE OF THIS BOOK

This book is the next stepping-stone in the roadmap described in the first book of the series (*Math Made a Bit Easier: Basic Math Explained in Plain English*, ISBN 1449565107). It covers key topics in basic algebra and geometry that commonly appear on end-of-grade exams, career-based exams, and exams such as the (P)SAT and GED. The book utilizes the same independent writing style as the first three books of the series.

PREREQUISITE KNOWLEDGE FOR THIS BOOK

This book assumes that you have *completely mastered* all of the material covered in the first book, and that you have practiced it using the practice exercises and self-tests in the second book. That statement is to be taken *quite literally*. This book moves at a very fast pace, and does not take the time to review concepts of basic math. If it did, the book would be twice as long, cost twice as much, and the flow of instruction would be interrupted.

If anything is not fully clear to you, ensure it is not because you forgot or never learned something which was covered in the first book such as how to compute -7 – (-3), or how to work with basic numeric fractions.

HOW THIS BOOK IS ORGANIZED

Unlike the first book of the series, this book does not include editorial content or tips on how to effectively study. Chapters One through Six cover topics in basic algebra, and Chapters Seven through Ten cover topics in basic geometry. The chapters should be read and mastered in order. Each topic builds upon previous topics, and geometry utilizes many concepts from algebra. Each chapter includes a small number of related practice exercises, as well as important points for review.

The book also includes an assessment test of the prerequisite basic math knowledge which is required to proceed with this book. It is nothing more than a variant of the tests that were included in the workbook companion to the first book in the series, but you should use it to determine if you need to review earlier material.

The book also includes a comprehensive end-of-book test, and concludes with information about contacting the author for help with the material presented.

THE BOOK'S POSITION ON CALCULATOR USE

As with the entire series, this book takes a realistic and modern position on calculator use. Feel free to use a calculator when needed while following along with the examples in this book. All of your effort should be to master the concepts being taught. If you do not fully understand a concept, not only will the use of a calculator not help you, it will almost certainly hinder you.

THE BOOK'S POSITION ON WORD PROBLEMS

As explained in the first book, it is inefficient to prepare for word problems on an exam by repeatedly reading and solving a few sample word problems in a book. The word problems you face on an exam will almost certainly be different, and changing even one word can drastically alter an entire problem.

The only way to prepare for word problems is to fully master all the topics on which you will be tested. Of course, it is essential that you become a skilled and careful reader, but that is something that cannot be learned from a book. It is something that must develop over time. It is also important to develop good test-taking skills in general as discussed in detail in the first book.

GRADE LEVEL OF THE MATERIAL COVERED

As described in the first book, it is actually this concern that leads to the struggle with math. Math is not about having a student in a particular grade do math which has been designated as being "on level" for that grade. Math must be learned step-by-step starting from the very beginning, progressing at whatever pace is necessary.

If you want to get better at math, stop thinking about grade levels, and stop thinking about tomorrow's test for which you may feel completely unprepared. Just think about learning math. Admittedly, it is hard to think in those terms when you are facing a particular academic deadline, but math will only start to get easier once you do.

WHY IS ALGEBRA HARD?

Most students are quick to say, "Algebra is hard," and in particular that they, "Never did quite get the whole thing with the x." The reality is that for most students, it is not the algebra that is the source of the problem. The issue is usually that the student never fully mastered the basic math topics upon which algebra is based.

As a simple example, if you are not completely sure of the answer to $3 - 5$, you will be even more unsure of how to simplify $3x^2 - 5x^2$. This is because doing so not only requires computing $3 - 5$, but also the knowledge of how to handle the x^2 terms that have been tacked on.

THE READER'S ASSUMED MATH EXPERIENCE

This book assumes that the reader has at least seen the material that is being presented, even if it was a long time ago and/or was never fully understood. It is extremely hard to learn brand new topics from any book, especially this one which attempts to include a large amount of material in relatively little space.

The truth is that basic algebra is actually quite easy because it just involves memorizing some basic concepts, and following some simple step-by-step procedures. Again, though, the problem usually comes down to the student not having fully mastered the basic math upon which the algebra is dependent. It is unfortunate that most students are "socially promoted" through the lower grades without being alerted to the abstract math which is in store for them in secondary school.

HOW TO USE THIS BOOK FOR SELF-STUDY

The advice in this section is the same as given in the first book. Unfortunately, many people will buy this book but will not use it. Some will not look at it at all, but will feel good that it is on their shelf. Some will quickly flip through it and say, "Yeah, I know all this stuff," when all they do is recognize the material or have a basic familiarity. Others may appreciate what this book has to offer, but will not have time to read it since they are scheduled for an exam in a few weeks if not days.

As obvious as this sounds, if you want to benefit from this book, you will have to read it. You will have to start at the beginning, and carefully read it to the end. You will need to be disciplined about not skipping over any chapters or sections that you feel you already know. You will need to go through the material as slowly and as often as you need to until you have fully grasped all of the concepts, and can perform related problems quickly, easily, and accurately on your own.

This book only includes a few sample problems and practice exercises for each topic. Your time and effort should be spent studying the actual concepts involved in each topic, instead of just copying the sample problems into your notebook over and over again. Doing that is an inefficient technique called "rote learning."

There are very few math topics for which learning by rote is beneficial or even sufficient. When solving a problem, it is important to think about what you are doing and what is going on. You must constantly focus on the connections between the different topics so you will learn how they are related, and not get them mixed up.

WORKING WITH THE SELF-TESTS AND EXERCISES

The book begins with an assessment test of prerequisite knowledge. It covers the basic math concepts that must be fully mastered if you are to have success with this book. Before proceeding with this book, it is essential that you feel fully comfortable with all of the questions on the assessment test. It is not sufficient to be satisfied with a "passing" grade of 65 since that will result in most of the material in this book being "hard" for you. Algebra and geometry build upon basic math concepts such as fractions, decimals, and signed number arithmetic.

If you are not finding the assessment test to be relatively "easy," put down this book, and read or reread the first two books in the series. If you cannot or choose not to buy them, you can read them for free on *Google Books*.

Each chapter concludes with a few related practice exercises. Avoid doing the exercises immediately after studying the chapter since that will not prove that you are truly retaining the information. As mentioned, the most important goal is to master the concepts presented. The way to do that is by thinking about the material, and how the concepts relate to one another. It is rarely of benefit to mechanically solve an endless list of practice exercises without any thought behind it. The book concludes with a comprehensive "final" exam comprised of variants of the chapter exercises. As with the practice exercises, allow some time to elapse between the time you finish the book and the time you attempt the exam. Don't engage in "spit-back" learning.

If you are finding yourself confused by any questions on the test, restudy the corresponding material. Learning math is a process that absolutely cannot be rushed.

SCOPE OF OTHER BOOKS IN THIS SERIES

This book is the fourth in a series of at least eight. The first book covers topics in basic math that form the foundation of all later material. The second book offers practice exercises and self-tests to serve as a companion to the first book. The third book is a companion book of lesson plans for teachers, tutors, parents, and homeschoolers.

This book begins the second "trilogy" of the series covering basic algebra and geometry. It was decided, however, for the companion workbook to be incorporated into this book. This is because at this level of math, there is simply not that much material to cover. The next book will be a companion book of lesson plans.

The third "trilogy" of the series will cover more advanced topics in algebra and geometry. Future books in the series may cover higher-level math such as pre-calculus and calculus, as well as very basic math that parents can teach to their young children.

Most students should find that the information in this and the first book is more than sufficient to fulfill the objectives of most math exams. It is worth noting that even graduate school entry exams such as the GMAT and GRE are only based on material that is approximately 10th grade level. However, those exams expect complete mastery of the concepts, and the ability to apply them in creative ways under extreme time pressure.

CHOOSING THE STARTING BOOK IN THE SERIES

No matter what your math level, it is *absolutely essential* that you begin with the first book in the series. As described in that book, the reason why math is hard for so many people is that they refuse to learn it from the beginning, and learn it thoroughly. Even if you are studying for a test in algebra or geometry, you should start with the first book. Even if you took a placement test which states that you are at an advanced grade level in math, you should start with the first book.

This is the path that you must take if you want to permanently and completely end the cycle of struggling with math. This series of books is a roadmap, and the road begins with the first book.

For students who cannot afford or do not wish to purchase the first book and its two companion books, they can be read for free in their entirety on *Google Books*. Visit www.MathWithLarry.com for details.

THE TARGET AUDIENCE FOR THIS BOOK

While this book does not target any particular audience, it is best suited for older students, adults, and parents of students. It should prove useful to students studying for their end-of-grade (EOG) or high school exit exam, college entrance exam, (P)SAT, GED, or similar exams such as career-based ones. The book should also prove valuable in a home-schooling environment. It may also be of benefit to teachers, including math teachers who are looking for a different perspective, as well as new teachers or those teaching math outside of their area of expertise due to a math teacher shortage situation.

WHAT THIS BOOK WILL *NOT* ACCOMPLISH

The purpose of this book is to help you get through whatever math requirements you are facing. It won't make you a math whiz, and it won't result in very high grades if you are very far behind in math, and have limited time in which to study. The book probably won't bring you to the point where you actually enjoy doing math. As the title suggests, after reading the book, you should find that math has been made a bit easier, and it should help you achieve your math goals.

This book is certainly not intended for anyone who is pursuing a math-based degree or career. Such students must study math from full-fledged textbooks as part of a highly enriched classroom setting.

Students who are pursuing math-based degrees or careers but require extensive math help should consider whether they have chosen the best path for themselves.

As discussed, this book will serve little to no benefit if you are scheduled to take an important exam tomorrow or next week, and are sorely unprepared for it. Flipping through the pages or sleeping with the book under your pillow is not going to help you. If you have bought this book as the result of such a "panic" situation, the limited time between now and your exam is best spent on mental/emotional preparation and learning test-taking techniques as opposed to last-minute academic cramming which will likely just fluster you even more. This is discussed more in the first book.

INTENTIONAL REDUNDANCY IN THE BOOK

As with the first book, this book is intentionally redundant to drive home certain key points. It is far better to present such concepts repeatedly and in different ways than to just mention them in passing. This is one of the biggest problems with traditional textbooks and review books. No author or publishing company wants to risk being accused of "going on and on" about the same thing. This book takes that risk. There is no such thing as over-learning very significant topics. Conversely, there is no point in spending a great deal of time on topics that most students either find very straightforward, or do not serve as important prerequisites (required prior knowledge) for later material.

A NOTE ABOUT ERRORS / TYPOS IN MATH BOOKS

Virtually all books go to print with some undiscovered errors or typos. Math books are especially prone to this. Unfortunately, many commercial publishers rush to get their books on the shelf before their competitors, resulting in an even greater number of errors. In the case of this book and all other books in the series, any errors discovered after publication will be noted and explained on my website. I also offer free copies of my books to anyone who informs me of a major error.

THE BOOK'S POSITION ON LEARNING DISABILITIES

This important section is copied verbatim from the first book. The subject of learning disabilities is extremely sensitive, and everyone has their own opinions on the matter. I firmly believe that while not everyone has the aptitude or interest to conquer college-level math, virtually everyone can learn to handle the math which is required to earn a high school diploma or similar. Some students must simply work harder than others to overcome whatever special challenges they may face. It is no different for math than for anything else in life. Every one of us has things that we find easy, and things that we find more challenging. We must work less on the former and more on the latter.

I also believe that a learning disability can sometimes become a self-fulfilling prophecy for some students. We are very quick to apply labels to people, and if the same labels are repeated often enough, we tend to integrate them into our sense of selves, and live up to them.

The path to learning math is exactly the same for every student, but each student will walk that path differently. Math must be learned thoroughly and step-by-step starting with the earliest material, and progressing systematically at whatever pace is required by the individual student. Topics that the student finds easy can be covered quickly, and topics that the student is struggling with require additional time and effort, but certainly cannot be rushed or skipped.

Rushing or skipping over topics sets the stage for struggling with math later on, at which time the math that was never learned in the first place must be learned. It is the proverbial "sweeping under the rug" which does not make the dirt actually go away.

The main reason why students struggle with math is because a typical school setting is not sufficient to teach math in the manner described above, and most parents cannot afford extensive and on-going private tutoring.

THE BOOK'S POSITION ON EXPERIMENTAL CURRICULA AND "NEW" MATH

This important section is copied almost verbatim from the first book. More and more schools now use unconventional math curricula. This is yet another matter that is the subject of controversy and debate. Some programs make extensive use of tangible manipulatives intended to aid students who are considered to be visual or kinesthetic learners. Other programs employ cooperative learning, group work, experiments, and student presentations, and others attempt to teach math by way of elaborate projects which in some cases are integrated with coursework from other subjects and/or "real-world" scenarios. Many programs utilize obscure and contrived algorithms (procedures) for solving simple problems.

This book explains math concepts in a very straightforward, traditional, conventional manner. It is important to understand that a typical standardized test such as the SAT, GED, college entrance exam, or career-based exam does not provide any leeway for the use of special physical props and/or special procedures. In fact, it is often the case that using such devices is even more complicated, time consuming, and error-prone than just solving a problem in the traditional manner. The exams are specifically designed to test whether or not the student can perform problems quickly and accurately.

As expected, the position of this book is that if you want to end the struggle with math, you must learn real math. I have personally observed countless students fail their exams while saying that the special devices and procedures used in class did not serve them well on their exam. It is critical to avoid taking a basic algebra topic such as factoring polynomials, and turning it into a huge production. While diagrams and tangible "algebra tiles" may have some merit as initial learning aids, at some point they must be set aside to transition to learning the basic procedure and all of its underlying concepts.

SO NOW WHAT?

Start by taking the assessment pretest to see if you have the required background knowledge to proceed with this book, or if you first need to review earlier material.

Chapter One introduces basic concepts of algebra starting from "the beginning." Do not rush through the chapter since it forms the foundation of the rest of the material presented in the book. Do not move ahead to Chapter Two until Chapter One has been mastered.

Assessment Pretest of Prerequisite Material

This self-test is comprised of variants of questions from the tests in the basic math workbook. Do not proceed with the material in this book unless you can complete these questions with confidence and ease. It is impossible to tackle algebra and geometry if you never learned or are not comfortable with concepts of basic math. Algebra and geometry build upon those concepts, just as stories of a building are constructed upon a solid foundation.

Take time to review the first book of the series and its companion workbook. Both are available for free reading in their entirety on *Google Books* if you cannot or do not want to purchase them.

Once you are fully comfortable with the basic math concepts that comprise this assessment test, proceed with the algebra and geometry content of this book. The self-test at the end can be used as either a pre-test, post-test, or both. All that matters is ensuring that you feel comfortable with the underlying concepts, but that is an assessment which only you can truly make.

1) Which basic operations are commutative?
2) Find the product of 9 and 7
3) Find the sum of -5 and 3
4) List the first 10 positive multiples of 8
5) Compute $68 \div 5$ in mixed number format
6) Insert "<" or ">": 65,999   66,001
7) Evaluate $5 \times [12 \div (-4 + 1)]$

8) Write "Seventeen billion, two hundred one million, eighteen thousand forty" as a number

9) Round 38,652 to the nearest hundred

10) Round 249,917 to the nearest thousand

11) Evaluate: $\sqrt{4}$. Include both roots.

12) Write "Seven hundred six and ninety-four hundredths" as a number

13) Insert "<" or ">": 5.18   4.99999

14) Insert "<" or ">": 0.89   0.8876

15) Convert to a fraction: 0.037

16) Convert 3/7 to a decimal (round to the nearest hundredth)

17) Convert to a decimal by hand: 6/50

18) Convert to a decimal: 2/3

19) True/False: 0.6 = .6 = 0.60

20) True/False: 0.3 is a terminating decimal.

21) Insert "<" or ">" (no calculator): 32/100   11/30

22) Insert "<" or ">" (use calculator): 540/802   431/642

23) Multiply (no calculator): 6.18 × 10,000

24) Divide (no calculator): 23.4 ÷ 1000

25) Round 23.4567 to the nearest thousandth

26) Round 59.9953 to the nearest hundredth

27) What is the reciprocal of 6/13?

28) Convert 13 ¼ to an improper fraction.

29) How is a meter related to a kilometer?

30) Apples are being sold at the rate of 29 apples for $14. How much does one apple cost at that rate?

31) Convert 62 feet to inches

32) Solve for the unknown value: $\frac{4}{9} = \frac{20}{?}$

33) Simplify to a single fraction: $13 / (\frac{2}{9})$

34) Find the mean of this list (rounded to the nearest integer): 36, 194, 78, 0, 756

35) Convert -6 to an equivalent fraction.

36) What is the GCF of 24 and 36?

37) What is the LCM of 4 and 6?

38) Reduce 23/37 to lowest terms

39) Define: Integer

40) Evaluate $(-4)^1$

41) Evaluate $\sqrt{49}$

42) Evaluate $2 + 5 \times -3$

43) Evaluate $-6 - 4 + 2$

44) Compute: $3 + (-8)$

45) Compute: $(-9) + (-8)$

46) Compute: $2 - 6$

47) Compute: $(-3) - 3$

48) Compute: $(-7) - (-2)$

49) Compute: $(-6) \times 9$

50) Compute $(-8) \times (-7)$

51) Compute: $10 \div (-2)$

52) Compute: $(-8) \div (-1)$

53) Evaluate: $(-16)^2$

54) Evaluate: $\sqrt{-25}$

55) Evaluate: 5^4

56) List the factors of 58

57) List the factors of 29

58) Compute: $0 \div -16$

59) Compute: $-12 \div 0$

60) Evaluate: (-3) squared

61) Multiply: $7 \times \frac{3}{11}$

62) Add: $\frac{1}{3} + \frac{1}{4}$

63) True/False: $\frac{14}{21} = \frac{140}{201}$

64) True/False: $\frac{13}{14} = \frac{15}{16}$

65) True/False: $-\frac{2}{7} = \frac{2}{7}$

66) True/False: $\frac{-4}{7} = \frac{4}{-7}$

67) Add: $\frac{4}{19} + \frac{5}{19}$

68) Multiply: $\frac{3}{8} \times \frac{4}{8}$

69) Divide: $\frac{7}{9} \div \frac{9}{7}$

70) Multiply: $\frac{234}{567} \times \frac{567}{234}$

CHAPTER ONE

Working with Algebraic Expressions

WHAT EXACTLY IS ALGEBRA?

Algebra is a vast topic of study in the field of mathematics. However, most students are only responsible for learning the very basics of the subject. At that introductory level, algebra is simply the study of how to solve for and work with unknown values.

In the lower grades, students work with problems that include unknown values such as $4 + \square = 5$, or $? - 2 = 6$. These problems are actually just simple, one-step algebraic equations, although they are not presented in those terms until much later. In earlier math, such problems are sometimes referred to as "number sentences." They are usually solved either intuitively, or by working backwards, or by using a guess-and-check method.

In later math, students go on to face problems that are essentially the same as the above samples. However, instead of the unknown values being represented by a box or a question mark, they will be represented by a letter of the alphabet, often x. The equations above would be represented as $4 + x = 5$, and $x - 2 = 6$. Instead of being solved using the methods described, they will be solved using specific procedures which lead to the answer even in problems that involve more complicated numbers.

x That is all algebra is at the basic level. It is the study of how to work with problems that contain one or more unknown values. Assuming you feel fully comfortable with all the math up to this point, the hardest part of algebra is just getting past the fear of "the x." In fact, algebra is actually easy because all problems can be solved using a simple, step-by-step procedure which will always lead to the correct answer.

NEW WAYS OF REPRESENTING MULTIPLICATION

In earlier math, we used the × symbol to represent multiplication. As mentioned, though, in algebra we often use the letter x to represent an unknown value. This means we must abandon the use of the × symbol for multiplication, since using it would be confusing.

In algebra, we have three different ways of representing multiplication. One way is by using what is called a **middle dot**. For example, we can write the equation $3 \cdot 5 = 15$ to mean "three times five equals fifteen." The middle dot is written higher than a decimal point as shown. When writing it by hand, ensure that it doesn't look like a decimal point or a stray mark on the page.

Another way to represent multiplication is by using parentheses. For example, we can write $(3)(5) = 15$. Remember that according to PEM-DAS, we would first evaluate whatever is in parentheses. In this case, there is just a value with nothing to actually evaluate. What we are left with is two values that are not connected by any operation symbol. In such a situation, we have what could be thought of as "implied multip-lication." The absence of any operational symbol tells us to multiply.

$$2x = 2 \cdot x = (2)(x) = 2 \times x$$

Three new ways of representing multiplication

Now that we know that the absence of an operational symbol means multiplication, there is one additional way of representing multiplication. If we write $3x$, what we actually mean is 3 times x. Some important points must be made. Remember, we are using x to represent some unknown value. It will make more sense shortly, but for now just accept that $3x$ means the value 3 multiplied by some unknown value denoted by x.

Do not get confused and think that $3x$ involves multiplication just because the letter x looks like the $\times$ symbol which was used in earlier material. We have a known value, 3, and an unknown value, x. They are adjacent to each other in the absence of any operational symbol. Recall that in such situations, the missing symbol implies multiplication. Just to be clear, the term $7y$ means 7 times y, since no operational symbol connects the 7 and the y.

EVALUATING ALGEBRAIC EXPRESSIONS

Very informally, an **algebraic expression** is a collection of numbers, letters, and operation symbols which doesn't include an equals sign. If such a collection did include an equals sign, it would be known as an **equation**.

Let's look at a very simple algebraic expression: $5 + x$. Translated into words it means, "The value 5 plus some unknown value called x." It is very important to understand that there is nothing to actually do with such an expression if it is presented in isolation. We are not told the value of x, and we can't deduce it since we are dealing with an expression and not an equation such as $5 + x = 8$.

The most common task involving an algebraic expression is to **evaluate** it using a stated value. For example, a problem may read, "Evaluate $5 + x$ when x is 2." We are given an expression, and are specifically told the value of the unknown quantity x. The person creating the problem was free to choose any value of x that s/he wanted. S/he happened to choose 2 for this particular problem. All we do is **substitute** the given value of x in the expression, and evaluate it. In this case, we get $5 + 2$ which we know is 7. Students sometimes informally refer to this procedure as "plugging in" the stated value of x.

Let's look at another problem: "Evaluate $10 - 2z$; $z = 3$." In this alternate but equivalent notation, the semicolon is being used in place of the word "when," and the equals sign is being used in place of the word "is." The problem also uses z instead of x, which we noted was fine.

In this problem, we are still dealing with an expression and not an equation. The equals sign is just telling us the value that the question-writer wanted z to take on for this particular problem. We have $10 - 2 \cdot 3$. Notice how the middle dot is being used to represent multiplication. All we must do is substitute the value of z in the expression, and evaluate according to PEMDAS. The expression evaluates to 4 which is our answer.

SOME IMPORTANT BASIC DEFINITIONS

Before continuing, it is important to become familiar with some basic algebra terminology. Since it is unlikely that you are required to know formal definitions, everything will be explained informally and by example.

We refer to an isolated numeric value as a **constant**. We refer to a letter representing an unknown value as a **variable**. The term "variable" can be confusing and a bit misleading. It is discussed in detail in Chapter Three. When we have a numeric value multiplied by a variable such as $2x$, we refer to the number as the **coefficient**.

We use the word **term** to mean any of the following: A constant, a variable, or the product or quotient of constants and variables. Examples of terms are 7, -10, x, w, $12z$, $a/3$, and $4b^2$. In the last example, what we actually have is a constant (functioning as a coefficient), times a variable, times that same variable again. **Expressions** with more than one term are formed by combining different terms using addition or subtraction.

SIMPLIFYING BY COMBINING LIKE TERMS

Many algebraic expressions can be **simplified** prior to evaluation. This is usually done by what is referred to as **combining like terms**. Let's look at an example. Evaluate $3x^2 + 7x + 4 + 6 + 5x^2 + 8x$ when x is 2. In plain English, all we do is combine everything that matches (i.e., the like terms). We will combine the constants of 4 and 6, the x terms of $7x$ and $8x$, and the x^2 terms of $3x^2$ and $5x^2$. Because of the properties of addition, it doesn't matter if the like terms are adjacent to each other.

$$\left(3x^2\right) + \boxed{7x} + 4 + 6 + \left(5x^2\right) + \boxed{8x}$$
$$\rightarrow \ 8x^2 + 15x + 10$$

Simplifying an algebraic equation by combining like terms

After combining, we have the simplified expression $8x^2 + 15x + 10$ which we were asked to evaluate when x is 2. Substitute 2 in for each occurrence of x, giving us $8(2)^2 + 15(2) + 10$. Evaluate according to the rules of PEMDAS to get 72. Notice that once we substituted a value in place of the variable, we used the parentheses notation for multiplication to avoid confusion. Again, this problem did not involve an equation. We were given an expression to evaluate, and were told what value to use for the variable. The question-maker could have chosen any value for x that s/he wanted.

If the same variable letter appears more than once in the same problem, it will take on the same value for each occurrence. In our problem, we substituted (plugged in) the value 2 everywhere that we saw an x.

MORE ABOUT COMBINING LIKE TERMS

Not every expression can be simplified. For example, in the expression $4x^2 + 4x + 4$, we have three completely different types of terms—a squared variable term, a non-squared variable term, and a constant. You could think of it as adding apples, bananas, and cherries. They cannot be combined. Resist the temptation to combine any terms that do not match. In this problem, it is just a coincidence that each term involves a 4. We are still dealing with different types of terms.

Let's evaluate the expression when $x = 5$. We can't do any simplification by combining like terms since there aren't any, so we'll just substitute and evaluate. We have $4(5)^2 + 4(5) + 4$ which evaluates to 124.

THE DEGREE OF A TERM

We define the **degree** of a term as the value of the exponent involved. For example, the term $9x^3$ is a third-degree term. The term $3x^2$ is a second-degree term. The term $8x$ is a first-degree term. Remember that x is an abbreviated version of x^1. The constant term 5 is actually of degree 0. To informally prove this, we could represent 5 as $5x^0$, remembering that any value to the power of 0 is equal to 1, and anything times 1 equals itself.

Remember: We are not allowed to combine terms whose degrees do not match.

THE IMPLIED COEFFICIENT OF 1

How can we simplify the expression $3x^2 + x^2 + 5x^2$? We have three like terms, all of second degree, so they can certainly be combined. All we must remember is that the middle term has an implied coefficient of 1. Some students like to call it an "invisible" 1. Remember that multiplying a value times 1 doesn't change the value, so x^2 is the same as $1x^2$. After combining like terms, the above expression simplifies to $9x^2$.

Remember: A missing coefficient is an implied 1, not 0. This is because a coefficient is meant to be multiplied by the variable that it precedes. It is not added to it.

IS IT SUBTRACTION OR A NEGATIVE SIGN?

Look at this expression: $2x^2 + 6x - 4 + 7 - 9x^2 - 8x$. It is common for students to look at the minus signs and ask, "How do I know if they are negative signs or subtraction signs?" It is important to understand that an expression like the one in the example is a series of terms connected via addition or subtraction. We can never have two terms sitting next to one another "in a vacuum" without being connected with addition or subtraction.

With that in mind, any minus signs that we see in the midst of an expression could be thought of as subtraction signs. However, when it comes to the task of combining like terms, we will, in a sense, treat those minus signs as negative signs. Here is the previous example once again: $2x^2 + 6x - 4 + 7 - 9x^2 - 8x$.

Let's start by combining the x^2 terms. There are two equivalent ways to proceed. If we treat the minus sign to the left of the $9x^2$ as a subtraction sign, we will be dealing with $2x^2 - 9x^2$ (2 minus 9) which is $-7x^2$. We could also choose to think about the minus sign to the left of the $9x^2$ as a negative sign. If we do, we will need to combine $2x^2$ and $-9x^2$, and combining in this context implies addition. That leads to $2x^2 + -9x^2$. To compute that, we would need to add $2 + (-9)$ which we know is -7. Either way, we end up with $-7x^2$.

$$\left(2x^2\right) + \boxed{6x} \left< - 4 \right> \left< + 7 \right> \left< - 9x^2 \right> \boxed{- 8x}$$

$$\rightarrow -7x^2 - 2x + 3$$

Simplifying an algebraic equation by combining like terms

Now let's combine the x terms. We can compute 6 – 8, or 6 + (-8). Either way, we end up with -2x. Combining the constants, we have -4 + 7, giving us 3. Our combined terms are $-7x^2$, -2x, and 3. Recall that we are not allowed to just write terms next to each other without being connected. An expression involves the sum or difference of terms—it isn't just a list of terms.

One way to represent our combined expression is by connecting the above terms with addition signs. That would give us $-7x^2 + -2x + 3$. Usually, though, we would just treat the minus sign to the left of the 2 as a subtraction sign, and write $-7x^2 - 2x + 3$.

Admittedly, it's a bit confusing, so reread this section slowly and carefully. The main point to understand is that subtracting a number is the same thing as adding its negative, as described in the first book of this series. For example, 3 – 5 is the same as 3 + (-5). Both equal -2.

PRACTICE EXERCISES AND REVIEW

Your most important task is to overcome any insecurity you may have about "The X." It is not an undiscovered 11th digit from another planet, nor something that your teacher or textbook invented to confuse you.

Remember: The variable x just represents an unknown value. Sometimes we are told what its value is, and then we must use that value in some way. Sometimes we must use special techniques to deduce what its value must be in order to make a given equation true. Remember, we can use any other letter instead of x.

Remember: If we are given an expression such as $2x + 7$, there is nothing to actually "do" with the x. If we are given a specific value of x, we can substitute it (i.e., "plug it in") and evaluate the expression, perhaps after first simplifying it by combining like terms if possible.

For practice, try these exercises:

1) What are four ways of representing "seven times z?"
2) Evaluate $-4x - 13$ when x is -3
3) Combine like terms: $-8x^2 - x - 2 + 4x^3 - 2x + 7x^2$
4) Combine like terms: $7x^3 + 5x^2 - 7x + 5$
5) Simplify: $a^3 + a^3 + a^3$

SO NOW WHAT?

In the next chapter, we'll work with algebraic equations—expressions that have been set equal to a value. Instead of being told the value of the variable to use like the examples in this chapter, we will be asked to determine what value of the variable will make the equation hold true. Before progressing to the next chapter, make sure you fully understand this one. We will apply these concepts throughout the entire book.

CHAPTER TWO

Solving Basic Algebraic Equations

WHAT IS AN ALGEBRAIC EQUATION?

An **algebraic equation** is essentially an expression that has been set equal to a value. An example is $3x + 2 = 17$. It is important to understand the difference between an equation and an expression. In the last chapter, we worked with expressions. We were given an expression and a value to use for the variable. The question-maker was free to choose any value for the variable that s/he wanted. We evaluated the expression after substituting the given value for each occurrence of the variable. In some cases, we first simplified the expression.

Working with an equation is different. Notice how in the equation above, we were not told what value to use for x, nor would the question-maker have been free to arbitrarily choose a value for x. Our job is to **solve** the equation to determine what value of x will make the equation hold true. The question is effectively, "Three times what number, plus two, will give us 17?" The directions for such a problem will likely read, "Solve for x."

For a simple equation like that, the answer of "$x = 5$" may "jump out at you," and is easily checked by substituting to see that the left side evaluates to 17. It is also not hard to get the answer using a guess-and-check method. A guess of 4 would prove to be too low. The left side would evaluate to 14. A guess of 6 would be too high because the left side would evaluate to 20. Despite how easy this seems, what we need is a set of procedures that will allow us to solve any equation regardless of how complicated it is or what values are involved. That is what basic algebra is all about.

GENERAL PROCEDURE FOR SOLVING EQUATIONS

Let's look at a simple algebraic equation: $x + 5 = 8$. Obviously, the answer is 3, but as mentioned, we need a formal process that will always work. The general procedure in algebra is that we look to see what is being done to the variable, and then we "undo" it. The goal is to somehow manipulate the equation so that the variable is by itself on one side. Once it is by itself, it is easy to see exactly what it is equal to, namely whatever value is on the other side of the equals sign.

In our example of $x + 5 = 8$, x could have been by itself on the left, but it isn't because 5 is being added to it. Recall that subtraction is the **inverse operation** of addition. If we were to subtract 5 from the left side, that would "undo" the 5 that is being added to the x. Informally, the +5 and the -5 would "cancel out."

At this point, you may have the following question: "What gives me the right to just take something away from one side of the equation?" The answer to that question leads us to the "golden rule" of algebraic equations, discussed next.

THE GOLDEN RULE OF ALGEBRAIC EQUATIONS

The "golden rule" of algebraic equations is that whatever we do to one side of an equation, we must do to the other. In our example, we subtracted 5 from the left side in an effort to get x by itself. That is fine, but we must now do the same thing to the right side. Think of an equation as an old-fashioned balance scale. You can do whatever you

want to one side of the scale, but if you want to keep it balanced, you must do the exact same thing to the other.

Getting back to our equation of $x + 5 = 8$, we decided to subtract 5 from each side. We now have $x + 5 - 5 = 8 - 5$. That simplifies to x = 3 which is our answer, as expected. Let's check it by substituting it back into the original equation. (3) + 5 = 8, and 8 = 8. Our answer of 3 "works" in the original equation.

EQUATIONS INVOLVING ADDITION

Let's solve another equation involving addition, this time using more common notation. Let's solve $x + 29 = 78$. As with our previous example, we will subtract 29 from both sides in an effort to get x by itself. Here is how we usually illustrate solving an equation such as this:

$$
\begin{array}{rcl}
x + 29 &=& 78 \\
-29 & & -29 \\
\hline
x & & = 49
\end{array}
\qquad
\begin{array}{l}
\text{CHECK:} \\
\begin{array}{rcl}
x + 29 &=& 78 \\
(49) + 29 &\overset{?}{=}& 78 \\
78 &\overset{\checkmark}{=}& 78
\end{array}
\end{array}
$$

Solving an algebraic equation involving addition

A few points must be made. First of all, when we subtract 29 on the left, don't get confused and think that it is being subtracted from both the x and the 29. As explained above, we're really dealing with $x + 29 - 29$ which is just x after we combine like terms.

Also, understand that even though it is conventional to cross out the +29 and the -29 and say that they "cancel," what is really happening is that we are computing $29 - 29$ which is 0, leaving us with $x + 0$. In the world of addition, we know that 0 doesn't change anything, so we can just omit it. The +29 and -29 don't just magically "disappear."

CHECKING YOUR ANSWERS

Most textbook exercises include the instruction to illustrate a "**check**" of your answer. As shown in the example, we first rewrite the original equation. We then substitute what we believe is the correct solution in place of x. We then do any computations and simplifications that are necessary to make sure that both sides of the equation are equal. If they are, it proves that the solution is correct. If they are not, it means that the solution is wrong, and you must determine where you made an error. Never just write an answer and say, "This is what I got after doing the algebra steps." Your answer must "work" in the original equation, otherwise it is incorrect.

EQUATIONS INVOLVING SUBTRACTION

We follow the same procedure to solve equations that involve subtraction, but we use addition to "undo" the subtraction that is being done to the x. Let's solve the equation $x - 37 = 46$. As shown below, we'll add 37 to each side to "undo" the fact that 37 is being subtracted from x. This results in x being left by itself on one side which is what we want. All other aspects of solving and checking the problem are exactly the same as what we did in the previous problem involving addition.

$$
\begin{array}{rcl}
x - 3\cancel{7} & = & 46 \\
+\ \cancel{3}7 & +37 & \\
\hline
x & = & 83
\end{array}
\qquad
\begin{array}{l}
\text{CHECK:} \\
x\ -\ 37\ =\ 46 \\
(83)\ -\ 37\ \overset{?}{=}\ 46 \\
46\ \overset{\checkmark}{=}\ 46
\end{array}
$$

Solving an algebraic equation involving subtraction

EQUATIONS INVOLVING MULTIPLICATION

We follow the same procedure to solve equations that involve multiplication, but we use division to "undo" the multiplication that is being done to the x. Recall that division and multiplication are inverse operations. Let's solve the equation $3x = 87$. As shown on the next page, we will divide each side by 3 to "undo" the fact that x is being multiplied by 3. This results in x being left by itself on one side which is exactly what we want. All other aspects of solving and checking the problem are exactly the same as what we did in our previous problems.

$$\frac{3x}{3} = \frac{87}{3}$$

$$1x = 29$$

$$\text{CHECK:}$$
$$3x = 87$$
$$3(29) \stackrel{?}{=} 87$$
$$87 \stackrel{\checkmark}{=} 87$$

Solving an algebraic equation involving multiplication

Notice how we represented the division using a fraction bar instead of the $\div$ sign. This makes it easier to visualize the "cancelling" of the threes. It is important to understand that the reason why the threes "cancelled" in this problem is different than the reason why values "cancelled" in our problems that involved addition and subtraction. In this problem, we were dealing with $3 \div 3$ which is 1. In the world of multiplication, 1 is the value that doesn't change things, just like 0 plays that role in addition and subtraction. After the threes "cancelled," we were really left with $1x$, but since that is the same as just plain x, we omitted the 1.

Sometimes students ask why the divisor of 3 only divides into the coefficient of $3x$ (i.e., the 3), and not into the x as well. One answer is that we don't distribute division over multiplication. If the numerator contained an expression such as $x + 5$ or $x - 7$, we would then have to distribute the divisor of 3 over both components of the numerator.

We'll work with problems like that in the next chapter. Another way of looking at the problem is that we are really dealing with $3 \div 3 \cdot x$. Once the threes "cancel out" and become 1, we're done with them.

EQUATIONS INVOLVING DIVISION

We follow the same procedure to solve equations that involve division, but we use multiplication to "undo" the division that is being done to the x. Let's solve the equation $x \div 2 = 18$. As shown below, we'll start by arranging the division in fraction format. Then we'll multiply both sides of the equation by 2 to "undo" the fact that x is being divided by 2. This results in x being left by itself on one side like we want. All other aspects of solving and checking the problem are the same as what we did in the last problem involving multiplication

$$\frac{x}{2} = 18 \qquad\qquad \text{CHECK:}$$

$$2 \cdot \frac{x}{2} = 18 \cdot 2 \qquad \frac{x}{2} = 18$$

$$1x = 36 \qquad\qquad \frac{(36)}{2} \overset{?}{=} 18$$

$$18 \overset{\checkmark}{=} 18$$

Solving an algebraic equation involving division

Sometimes students ask what gives us the right to cross off the twos as shown. When we multiply the left side by 2, we're really multiplying by 2/1. Remember that when we're working with a problem that involves fractions, it is usually best to convert all integers to fractions. With that in mind, the multiplier of 2 is really over an "imaginary" 1, therefore "pushing" it up to the numerator. In that position, we are permitted to "cross cancel" as described in the first book.

EQUATIONS INVOLVING ADDING OR SUBTRACTING FRACTIONS

All of the techniques that we worked with can be used to solve algebraic equations involving fractions. The procedures do not even need to be changed at all, but you certainly must feel fully comfortable doing operations with fractions in a non-algebra context. If you don't, review the related material in the first book.

Let's try solving $(1/3) + x = (1/2)$. Parentheses have been added for clarity. The procedure is no different than solving $x + 5 = 8$ like we did. We must undo whatever is being done to the x so that it will be alone on one side of the equation. Whatever we do to one side must also be done to the other. In this case, since $(1/3)$ is being added to x, we must subtract $(1/3)$ from both sides. That gives us $x = (1/2) - (1/3)$, and after simplifying we have $x = 1/6$. Check the answer by substituting $(1/6)$ in place of x in the original equation. We have $(1/3) + (1/6) = (1/2)$. After simplifying and reducing, we can see that both sides do equal $(1/2)$ which proves our answer is correct.

$$\begin{array}{rcl} x + \frac{1}{3} &=& \frac{1}{2} \\ - \frac{1}{3} & & -\frac{1}{3} \\ \hline x & & = \frac{1}{6} \end{array}$$

$$\text{CHECK:}$$
$$\begin{array}{rcl} x + \frac{1}{3} &=& \frac{1}{2} \\ (\frac{1}{6}) + \frac{1}{3} &\overset{?}{=}& \frac{1}{2} \\ \frac{1}{2} &\overset{\checkmark}{=}& \frac{1}{2} \end{array}$$

Solving an algebraic equation involving adding fractions

Note that had the equation involved subtracting a fraction instead of adding one, we would follow the same procedure that we used for solving an equation that involved subtracting an integer. In the next chapter, we'll solve problems that involve multiplying the variable by a fraction. Problems that involve dividing the variable by a fraction are rare, and are not covered.

EQUATIONS INVOLVING ADDING OR SUBTRACTING DECIMALS

All of the techniques that we worked with can be used to solve algebraic equations involving decimals. The procedures do not even need to be changed at all, but you certainly must feel fully comfortable doing operations with decimals in a non-algebra context. If you don't, review the related material in the first book.

Let's try solving $x - 2.7 = 8.51$. The procedure is no different than solving $x - 37 = 46$ like we did. We must undo whatever is being done to the x so that it will be alone on one side of the equation. Whatever we do to one side must also be done to the other. In this case, since 2.7 is being subtracted from x, we must add 2.7 to both sides. That gives us $x = 8.51 + 2.7$, and after simplifying we have $x = 11.21$. The problem and its check are shown next.

$$
\begin{array}{ll}
\begin{array}{r}
x - 2.7 = 8.51 \\
+\ 2.7 \quad +2.7 \\
\hline
x \qquad\quad = 11.21
\end{array}
&
\begin{array}{l}
\text{CHECK:} \\
x - 2.7 = 8.51 \\
(11.21) - 2.7 \overset{?}{=} 8.51 \\
8.51 \overset{\checkmark}{=} 8.51
\end{array}
\end{array}
$$

Solving an algebraic equation involving decimal subtraction

PRACTICE EXERCISES AND REVIEW

In this chapter, we worked with algebraic equations Recall that equations involve an equals sign whereas expressions do not. In an equation, we are not just free to choose any value we want for x. We must determine the value of x that makes the equation true. For example, in the equation $x + 2 = 5$, x must take on the value of 3.

Remember: Unless the solution to an equation is obvious, we must use algebra to solve it. The golden rule is to "undo" whatever is being done to the x, with the goal of getting x by itself on one side. Anything we do to one side of the equation must also be done to the other.

40

Remember: Always check your answer by substituting it back in the original equation, and evaluating to see if both sides are equal.

For practice, try these exercises:

1) Solve for x: $x + 7 = -13$
2) Solve for x: $x - 4 = 12$
3) Solve for x: $-4x = 20$
4) Solve for x: $8x = 4$
5) Solve for x: $\frac{x}{-3} = -3$
6) Solve for x: $\frac{3}{5}x = 60$
7) Solve for x: $\frac{3}{8} + x = \frac{15}{16}$
8) Solve for x: $x - 4.6 = 7.13$

SO NOW WHAT?

Before progressing to the next chapter, it is essential that you fully understand all of the concepts in this one. Take as much time as necessary to review the material.

CHAPTER THREE

More Complicated Algebraic Equations

WHAT EXACTLY IS A VARIABLE?

$\mathcal{X}$ An unknown value such as the x in $3x + 7 = 12$ is often referred to as a **variable**. This can be a bit of a misnomer in some cases. In an equation, the variable does not actually vary. In fact, our task is to compute the value of the variable that will make the equation true, namely that both sides are equal. In basic equations there will be only one such value, and we are certainly not free to just choose it as we see fit.

That is not the case with variables in expressions like those that we worked with in Chapter One. An expression is not set equal to any particular value. For example, consider the expression $6x - 2$. There is nothing to actually do with an expression on its own other than possibly combining like terms which is not applicable in this case. In an expression, we are free to choose any value that we want the variable to take on, and we can then evaluate the expression using that value. In that sense, the variable does vary since we could evaluate the expression repeatedly using different selected values for the variable.

To review, if the same variable letter is used multiple times throughout the same problem such as $7x = 8 + 5x$, it will take on the same value throughout the entire problem. It may be a value chosen by us (or the

43

question-maker) in an expression, or it may be a value that we must solve for in an equation.

If the same variable letter appears in different problems, it may take on a different value in each problem, but it certainly doesn't have to. For example, it could just be a coincidence that x happened to equal 3 in two consecutive equations.

Remember, there is nothing special about the letter x as a variable. It is the most commonly used letter for that purpose, but any letter could be used in its place.

SIMPLIFYING BASIC ALGEBRAIC EQUATIONS

Some algebraic equations can be simplified before solving. All we do is follow the same procedures that we used in Chapter One to simplify algebraic expressions. Look at this equation: $-2x + 3x + 7x = 7 + 21 - 8$. All of the x terms are on the left, and all of the constants are on the right. This makes our job easier. After combining like terms on both sides, we have $8x = 20$. Divide both sides by 8 like we learned in the last chapter to get $x = 2.5$. Don't be alarmed if you ever get a fractional answer—just check it as we've learned.

$$-2x + 3x + 7x = 7 + 21 - 8$$
$$8x = 20$$

Combining like terms on each side of an algebraic equation

It is important to understand that we can only combine like terms that are on the same side of the equals sign. For example, in the equation $4x + 5 = 7x - 3$, we cannot do any combining since the x terms are on opposite sides, as is the case with the constant terms. Later in the chapter, we'll learn how to handle such situations.

SOLVING TWO-STEP ALGEBRAIC EQUATIONS

In the last chapter, we worked with what are known as one-step equations. They could all be solved in one step, namely "undoing" whatever was being done to the x, followed by simplifying and checking. However, many equations require two-steps to solve.

Look at this equation: $2x + 3 = 11$. Remember that the general procedure for solving equations is to get the x on one side by itself. We do this by using inverse operations to "undo" whatever we see being done to the x. Remember that whatever we do on one side of the equation we must also do on the other to maintain the balance.

In this example, we only have an x term on one side of the equals sign which is good. However, we do have constants on both sides. Our goal is leave our x term on the left since there is no reason to move it anywhere, and to move all of our constant terms to the right where they can be combined. Once we do all that, we'll be left with a simple one-step equation.

It may be tempting to begin by dividing both sides by 2, seeing that x is being multiplied by 2. While it is not wrong to do this, it is not the best way to proceed. The reason is that if we want to divide the left side by 2, we would not only need to divide the $2x$ term by 2 (cancelling the twos), but we would also need to divide the 3 by 2, resulting in a fraction of 3/2. This is because the left side contains an expression as opposed to a single term. After dividing the right side by 2, we would have 11/2 on the right. The truth is that we could still proceed and get the correct answer, but it will involve unnecessary work with fractions.

The best way to solve our equation of $2x + 3 = 11$ is to first "undo" the addition of 3 on the left. We know we can accomplish that by subtracting 3 on both sides. We have $2x + 3 - 3 = 11 - 3$. After combining like terms, we have $2x = 8$ which is much simpler than what we started with. Just divide both sides by 2 to get $x = 4$. To check, substitute 4 back into the original equation to see that both sides equal 11.

$$2x + 3 = 11$$
$$-3 \quad -3$$
$$\overline{2x + 0 = 8}$$
$$\frac{}{2} \qquad \frac{}{2}$$
$$x = 4$$

CHECK:
$$2x + 3 = 11$$
$$2(4) + 3 \overset{?}{=} 78$$
$$8 + 3 \overset{?}{=} 11$$
$$11 \overset{\checkmark}{=} 11$$

Solving a two-step algebraic equation

This example is by far the most common type of two-step equation that you will encounter. Just remember to first "undo" the addition or subtraction by performing the inverse operation on both sides, and then divide both sides by whatever value x is being multiplied by.

EQUATIONS WITH VARIABLE TERMS AND CONSTANTS ON BOTH SIDES

Some equations have variable terms on both sides of the equals sign. An example is $7x - 8 = 3x + 4$. As mentioned, we can't combine like terms across the equals sign. Informally, what we do in such situations is "move over" all of the x terms to one side, and "move over" all of the constant terms to the other. It doesn't matter which side ends up with which, as long as the all the variables and constants are on opposite sides. Then we can combine like terms and solve the equation as usual.

In this example, the first decision to make is which side will get the variable terms, and which side will get the constants. There is no wrong answer to that matter. However, it will be slightly easier if we don't end up with negatives on either side. Let's try moving the $7x$ term out of the left side and onto the right. We would need to subtract $7x$ from both sides. On the right we would have $3x - 7x$, which would give us $-4x$. While there is nothing wrong with that, we can avoid dealing with the negative by choosing to gather our variable terms on the left. We'll also avoid having a negative constant by moving all of our constants onto the right.

Let's start by subtracting $3x$ from each side. We then have $4x - 8 = 4$. Already that looks a bit better. Now that our variable terms are all combined on the left, we must get all of our constants on the right. We'll add 8 to both sides, giving us $4x = 12$. Now we have a simple one-step equation. Divide both sides by 4 to get $x = 3$.

$$
\begin{array}{rcl}
7x - 8 &=& 3x + 4 \\
-3x && -3x \\
\hline
4x - 8 &=& 4 \\
+8 && +8 \\
\hline
4x &=& 12 \\
&& \overline{4} \\
x &=& 3
\end{array}
$$

CHECK:
$$
\begin{array}{rcl}
7x - 8 &=& 3x + 4 \\
7(3) - 8 &\overset{?}{=}& 3(3) + 4 \\
21 - 8 &\overset{?}{=}& 9 + 4 \\
13 &\overset{\checkmark}{=}& 13
\end{array}
$$

An equation with variables and constants on both sides

Be sure to understand that we could have chosen to move all of our variable terms onto the right, and our constants onto the left. Try it for practice. You should end up with $-12 = -4x$. Divide both sides by -4 to get the same answer of $x = 3$. Be sure to correctly apply the rules for dividing signed numbers.

FINDING THE UNKNOWN VALUE WHICH RESULTS IN A SPECIFIED MEAN

A popular test question involves being given some test scores that a student received, and being asked to determine what grade the student would need to receive on his/her next test in order to have a certain average (mean). An example is as follows: A student received test grades of 63, 74, 89, and 94. What grade would the student need to earn on his/her next test in order to have a mean of 85?

We know that to compute the mean, we must add up all of the in-volved scores, and divide by the number of scores. In this case, we know that there will be 5 scores after the student takes his/her fifth test, but we don't know what that grade well be, so we'll call it x. We also know that we want the average to be 85, so that is what we'll need to set our equation equal to.

$$\frac{63 + 74 + 89 + 94 + x}{5} = 85$$

Determining the unknown value in an averaging problem

There is no harm in combining the constants in the numerator, and after doing so we have $(320 + x) / 5 = 85$. At this point, let's eliminate the denominator of 5 which is complicating

$$\frac{320 + x}{5} = 85$$

things. We do this by multiplying both sides of the equation by 5. As we saw in the last chapter, when we multiply by 5 we're really multiplying by 5/1, "pushing" the 5 up into the numerator which "cancels" the 5 in the denominator.

$$\cancel{5} \cdot \frac{(320 + x)}{\cancel{5}} = 85 \cdot 5$$

$$320 + x = 425 \quad \rightarrow \quad x = 105$$

**Simplifying to determine the unknown value in
a problem involving the mean of test scores**

Don't get confused and think that the 5 should be distributed over the 320 plus the x. In other contexts, we will do things like that, but in this problem, the fives will "cancel" before they have a chance to do anything else. Remember that the fives don't just vanish. We really have 5/5 which is 1, and we know that we don't need to ever show that something is being multiplied by 1. Be sure to notice that both sides were multiplied by 5.

After performing our computations and the basic algebra, we get $x = 105$. The student will need to receive a grade of 105 in order to have a five-test average of 85. With luck, there will be the opportunity to earn a bit of extra credit! Notice how one or two lower grades can really make it hard to bring up one's average.

EQUATIONS WITH A VARIABLE TIMES A FRACTION

A common equation format involves a fraction times a variable. These problems are easy as long as you remember the simplest way of solving them. To solve an equation like $4x = 12$, we "undid" the multiplication by dividing both sides by 4. We could certainly apply that same technique to an equation involving a fraction times a variable such as

$$\frac{2}{3}x = 14$$

the one at left. However, that would require us to divide both sides by a fraction. Recall that the first book presented a four-step process for dividing by a fraction. While it is not hard, it is four steps that must be remembered and carefully performed.

The simplest way to solve an equation such as the one in our example is to multiply both sides by the reciprocal of the fraction. Recall from the first book that multiplying a fraction times its reciprocal yields a product of 1. Informally, we can say that everything "cancels."

$$\frac{\cancel{3}}{\cancel{2}} \cdot \frac{\cancel{2}}{\cancel{3}}x = 14 \cdot \frac{3}{2} \quad \rightarrow \quad x = 21$$

Multiplying by the reciprocal to solve an algebraic equation
involving a fraction times a variable

THE SIMPLE INTEREST FORMULA

A popular test question involves what is known as the **simple interest formula**: **I = Prt**. P stands for principal. Think of it as the amount of money you use to open a bank account or begin an investment. The *r* stands for rate. It means the rate of interest your money will earn while it's invested. The reason why it's called the simple interest formula is that we won't worry about matters of compounding or anything of the kind. You will either be told what rate to use (usually presented as a percent), or you will need to compute it. You could think of it as the annual percentage yield (APY/APR), but you won't need to do any financial calculations.

The *t* stands for time, but it is very important to understand that this variable must always be *in terms of years*. This means that if the money is going to be invested for 6 months (i.e., ½ a year), you will need to use a value of 0.5 for *t*, and not 6.

The I stands for the amount of interest your initial deposit (the principal) will earn at a given rate (*r*), after a given amount of time *in years* (*t*).

$$I = Prt$$

Interest equals Principal times Rate times Time (in years)

In a problem involving the simple interest formula, you will be presented with three out of the four pieces of information. You must use algebra to compute the missing value which will not always be interest.

Let's look at an example: You are going to invest $3000 in a two-year CD that offers a rate of 4.9%. How much interest will you earn? In this case, we are given P, r, and t, and we must compute the value of I. Substituting, we have I = 3000(0.049)(2). Notice that we converted the percent to a decimal since we need to compute with it. After calculating, we get I = $294 which is our answer. If we were asked how much money the bank would owe us in total after the two years, we would of course have to add back in the original principal of $3000.

Let's look at one more example. You have $20,000 to invest, and you would like it to earn at least $800 in interest after having been invested for 9 months. What is the minimum interest rate that will result in such earnings?

In this case, our formula is 800 = 20000(r)(0.75). We are told the interest that we want to earn, as well as all other pieces of information except for the rate. Notice how we used 0.75 for the time since 9 months is ¾ of a year, and t must always be measured in terms of years.

Simplifying, we have 800 = 15000r, and solving algebraically for r we get r = 0.05$\overline{3}$ or 5.33% (rounded). We'll need to find an investment which offers at least that rate in order to achieve our interest goal.

PROPORTION PROBLEMS INVOLVING ALGEBRA

In the first book, we worked with proportion problems that were solvable without using algebra. In this section, we will learn how to solve proportion problems that do not have an obvious solution.

$$\frac{4}{7} = \frac{x}{10}$$ Recall that in a proportion, the cross products are equal to one another. To solve the proportion at left, begin by equating the cross products as 7x = 40. Divide both sides by 7 to get x = 40/7 or 5.7 (rounded). Let's informally check to see

if that answer makes sense. 4 is slightly more than half of 7. It therefore makes sense that x should be slightly more than half of 10, which it is. If you get an answer which doesn't fit that description, it is certainly wrong, and means that you made a mistake.

Remember: Any proportion problem with an unknown can be solved in this manner. Just equate the cross products and solve algebraically.

USING THE DISTRIBUTIVE PROPERTY TO ELIMINATE PARENTHESES

Algebraic equations often include an expression in parentheses which is multiplied by a term. In those situations, we must use the distributive property to remove the parentheses before solving.

Remember: The distributive property of multiplication over addition can be represented as $a(b + c) = ab + ac$. The formula also applies to subtraction: $a(b - c) = ab - ac$.

In the first book, we noted that the distributive property didn't seem very useful in situations where it was easy to just add or subtract whatever was in parentheses. In algebra, though we will often work with expressions such as $3(x + 2)$ where it isn't possible to combine what is in parentheses. The only way to remove the parentheses is if we distribute the term that is outside.

Let's look at the equation $3(x + 2) = -15$. The truth is that there are two different ways to solve this equation, one of which is to start by dividing both sides by 3. While that will work in equations such as this, it will not be that simple in other equations that look quite similar.

$$3(x + 2) = -15 \qquad \text{Check:}$$
$$3x + 2 \cdot 3 = -15 \qquad 3(x + 2) = -15$$
$$3x + 6 = -15 \qquad 3(-7 + 2) = -15$$
$$3x = -21 \qquad 3(-5) = -15$$
$$x = -7 \qquad -15 = -15$$

Using the distributive property of multiplication over addition to remove the parentheses in an algebraic equation

The best way to proceed is to first use the use the distributive property to remove the parentheses. Once that is done, we will be able to proceed just as we did in all of our previous examples.

Notice how we started by distributing the 3 over the x and the 2 in the parentheses. Once that was done, we were left with a basic two-step equation no different than we saw in previous examples.

Remember: When an algebraic equation includes a term that multiplies an expression in parentheses, it is usually best to first use the distributive property to remove the parentheses before proceeding.

Let's use the distributive property to remove the parentheses in the expression $-5(x - 7)$. The recommended method is

$$-5(x - 7)$$
$$-5x - (-5 \cdot 7)$$
$$-5x - (-35)$$
$$-5x + 35$$

to follow the pattern of $a(b - c) = ab - ac$. In this case, our a term is -5, our b term is x, and our c term is 7. The problem is illustrated at left. Notice how the -5 first multiplies the x, and then it multiplies the 7. Those two products are connected by subtraction. The hardest part of handling a problem like this is remembering all the rules for signed number arithmetic. If necessary, review the first book for details.

Understand that while it is possible to treat the minus sign before the 7 as a negative sign for the 7, if we do, we will no longer have an operation connecting the x and the -7. We are allowed to put in an implied addition sign between them, but all of that is unnecessary work if you proceed as described above.

PRACTICE EXERCISES AND REVIEW

This chapter covered a lot of material. Remember, before solving an algebraic equation, check to see if any combining of like terms can be done. Handle each side of the equation separately when combining.

No matter how complicated an equation is, all of the constants will have to be "moved over" and combined on one side, and all of the variable terms will have to be "moved over" and combined on the other side. It does not matter which side ends up with which. You will get the same answer in either case.

An example of a common equation form is $3x + 5 = 17$. The best way to proceed is to first subtract 5 from each side, giving us $3x = 12$. Then divide both sides by 3 to get $x = 4$. Be very careful with your signed number arithmetic in any equation that involves negatives.

Review the procedure for solving a mean (average) problem involving an unknown value. The process is always the same.

Remember: In equations involving a variable times a fraction, the best way to proceed is to multiply both sides by the fraction's reciprocal.

Remember: In the simple interest formula, time must stated be in terms of years. That means that 6 months must be represented in the formula as 0.5. Be careful to determine if a problem is asking about just the interest, or the interest with the principal added back in. Remember to convert the interest rate percent into a decimal.

Remember: In a proportion problem involving a variable, set the cross products equal to each other, and solve the resulting one-step equation.

Remember: The distributive property can be used to eliminate parentheses in expressions of the form $a(b + c)$, even in expressions that involve variables. When distributing a negative value, be certain that you are properly adhering to the rules of signed number multiplication. Also, remember that an expression such as $x - 5$ could be rewritten and treated as the equivalent $x + (-5)$.

<u>For practice, try these exercises:</u>

1) Solve for x: $-8 - (-7) + 4 = 7x - 9x + 12x$
2) Solve for x: $-2x - 9 = 15$
3) Solve for x: $-9x - 3 = -13x + 5$
4) A student received grades of 82, 96, 81, 42, and 63. What grade must s/he get on the next test to have a mean of 75?
5) A person invests \$3600 for 6 months in an account that offers an APY of 2.50%. How much money in total will s/he have at the end of the time period?
6) A person wants to earn \$475 on a \$4100 investment. If s/he is willing to invest the money for 3 years, what is the minimum interest rate s/he must find?
7) Solve for x: $\frac{3}{11} = \frac{13}{x}$
8) Solve for x: $5(x + 6) = 42$
9) Solve for x: $-10(x - 4) = -15$

SO NOW WHAT?

Before progressing to the next chapter, it is essential to fully understand all the concepts in this one. If you do not, you will have tremendous difficulty with later work, and will need to return to this chapter.

CHAPTER FOUR

Exponents in Algebra

UNDERSTANDING THE RULES FOR EXPONENTS

Exponents play a large role in algebra. If you are not fully comfortable with what an exponent is, and the rules for exponents of 0 and 1, and how to do basic computations with exponents, review the corresponding material in the first book before proceeding with this chapter.

This chapter presents some rules for how to handle exponents in algebraic expressions and equations. It is very easy to confuse one rule with another. Instead of trying to memorize the rules, it is far better to understand how they work. In doing so, you will be able to recreate a rule anytime that you need to.

PROBLEMS OF THE FORM $x^a \cdot x^b$

The first rule tells us how to multiply two exponential terms that have the same base, but different exponents, for example, $x^2 \cdot x^3$. The rule is that in situations like this, we add the exponents, and keep the base. The example simplifies to x^5. As mentioned, though, rather than memorize the rule, it is far better to understand it.

Think about what x^2 means. It means $x \cdot x$. We also know that x^3 means $x \cdot x \cdot x$. What happens when we multiply those two results? We have $x \cdot x \cdot x \cdot x \cdot x$, which is x^5. In this expression, we really just needed to count how many copies of x would end up being multiplied together.

$$x^a \cdot x^b = x^{(a+b)}$$

When we multiply two like bases, we add the exponents

Let's look at a simple example using actual numbers. Simplify: $12^4 \cdot 12^5$. Following the rule, the answer would be 12^9, which is how you would leave your answer.

"TRICK" PROBLEMS OF THE FORM $x^a + x^b$

It is important to understand that the previous section involved multiplying and not adding exponential terms with like bases. There is no formula for problems of the form $x^a + x^b$ in which we are adding like bases.

Let's look at an example: Simplify $2^4 + 2^8$. If we compute the values of each term, we get $16 + 256 = 272$. There is no simple formula involving the exponents that would get us to that value. In particular, the answer is not equal to 2^{12} which equals 4096. Don't get tricked by problems of this form. If you are expected to get a numerical answer, for now you will need to compute it as we did in the example.

PROBLEMS OF THE FORM $x^a \div x^b$

The next rule we'll look at is how to divide two exponential terms that have the same base, but different exponents. An example is $x^5 \div x^3$. The rule is that in situations like this, we subtract the exponents and keep the base. The example simplifies to x^2. As mentioned, though, rather than memorize the rule, it is far better to understand it.

Think about what x^5 means. It means $x \cdot x \cdot x \cdot x \cdot x$. We also know that x^3 means $x \cdot x \cdot x$. What happens when we set up the problem as a fraction, and divide the first result by the second, and "cancel" pairs of x's which appear both the numerator and denominator?

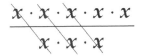

"Cancelling" pairs of common factors to see what remains

Remember that when we "cancel" common factors in a fraction, those factors don't just disappear. Anything of the form x/x can be reduced to $1/1$, which is further simplified to 1, and in multiplication, 1 is the identity element or "invisible" number which we need not write.

In this example, all three x's in the denominator are paired up with three of the x's in the numerator. In the denominator we are left with our implied 1 which we can ignore, and in the numerator we are left with $x \cdot x$. The problem of $x^5 \div x^3$ simplifies to x^2.

In this expression, we really just needed to count how many copies of x would remain after matching pairs in the numerator and denominator were removed. Notice that what we really did was subtract in order to find the difference between the number of x's in the numerator and the number of x's in the denominator.

$$x^a \div x^b = x^{(a-b)}$$

When we divide two like bases, we subtract the exponents

PROBLEMS OF THE FORM $(x^a)^b$

The next rule we'll look at is how to handle an exponential term which itself is raised to a power. An example is $(x^2)^3$. The rule is that in situations like this, we multiply the exponents and keep the base. The expression above simplifies to x^6. As mentioned, though, rather than memorize the rule, it is far better to understand it.

Remember that x^2 means $x \cdot x$. We also know that when we raise a quantity to the power of 3, it means to multiply it by itself so that we have three copies. What happens when we take $(x \cdot x)$ and multiply it times itself for a total of three copies? We have $(x \cdot x) \cdot (x \cdot x) \cdot (x \cdot x)$. We can remove the parentheses that were only added for clarity, and what is left is just x multiplied by itself for a total of 6 copies which is x^6. In this expression, we really just needed to count how many copies of x would end up being multiplied together.

$$(x^a)^b = x^{ab}$$

**When we raise an exponent term to a given power,
we multiply the exponents**

Remember: In all problems involving exponents, a variable by itself such as x should be treated as x^1.

PROBLEMS OF THE FORM x^{-a}

The next rule we'll look at is how to handle a variable raised to a negative exponent. The rule is that we make the exponent positive, and then put that value under a numerator of 1. While it is easy to remember that rule, it is beneficial to have a sense of where it comes from.

Let's look at the problem $x^3 \div x^5$. We can set it up as shown below. Notice how we can "cancel" pairs of x's as we did in an earlier problem. On top, we are just left with an implied 1, and on the bottom, we are left with x^2. This shows that $x^3 \div x^5 = 1/x^2$.

$$\frac{x \cdot x \cdot x}{x \cdot x \cdot x \cdot x \cdot x}$$

"Cancelling" pairs of common factors to see what remains

With that in mind, remember that we could have solved this problem using the formula that tells us to subtract the exponents since we are dividing like bases. If we did that (in the proper order), we would have x^{3-5} or x^{-2}. Informally, this shows that x^{-2} is the same as $1/x^2$, and of course the pattern holds for any negative exponent.

$$x^{-a} = \frac{1}{x^a}$$

When a value is raised to a negative exponent, make the exponent positive, and put the value under a numerator of 1

SO WHY DOES x^0 = 1?

In the first book, we said that any value raised to the power of 0 equals 1, but without explanation. Now that we're learning about how exponents work in algebra, an informal proof by example can be provided.

Let's look at the problem $x^3 \div x^3$. We are dealing with the situation of a quantity divided by itself which we know is always equal to 1. With that in mind, remember that we could have solved this problem using the formula that tells us to subtract the exponents since we are dividing like bases. If we did that, we would have x^{3-3} or x^0. Informally, this shows that x^0 is equal to any problem of the form $x^a \div x^a$ which in turn is equal to 1.

MULTIPLYING TERMS WITH COEFFICIENTS AND DIFFERENT VARIABLES

Let's look at an example that may seem complicated at first. Simplify: $7xy^2 \cdot 4x^3y^5$. This entire expression is nothing but a long chain of terms being multiplied together. The middle dot is being used in the center, and all of the other components are connected with no operation symbol which implies multiplication.

Remember that multiplication is commutative which means we can reorder terms as we see fit. It doesn't matter that the coefficients and matching variables are not actually adjacent to each other, although for now, it will be helpful to rearrange the problem so that they are.

Let's rearrange the expression, and insert parentheses and middle dots for clarity, but again, this is something that can be done mentally. We now have $(7 \cdot 4) \cdot (x \cdot x^3) \cdot (y^2 \cdot y^5)$. After multiplying the coefficients to get 28, all we must do is follow the rule for multiplying terms with like bases. After multiplying the x terms, we have x^4, and after multiplying the y terms, we have y^7. The original expression is most concisely simplified as $28x^4y^7$.

$$7xy^2 \cdot 4x^3y^5 = (7 \cdot 4) \cdot (x \cdot x^3) \cdot (y^2 \cdot y^5)$$
$$= 28x^4y^7$$

Multiplying terms with coefficients and different variables

Remember: If an algebra problem involves more than one variable letter such as x and y, each letter can be thought of as its own entity.

THE DISTRIBUTIVE PROPERTY REVISITED

In Chapter Two, we looked at the basic way in which the distributive property is used in algebra. Now let's look at some more challenging situations that arise.

Let's simplify $3x(5x + 7)$. Remember that we must distribute the $3x$ onto each term in parentheses. Our intermediate step is to simplify $(3x \cdot 5x) + (3x \cdot 7)$. Just follow all of the rules that we've been working with to get $15x^2 + 21x$. Review all of the previous material if you are confused about any aspect of that solution.

$$3x(5x + 7) = 15x^2 + 21x$$

Distributing a variable with a coefficient

Let's try another. Simplify $-2x(4x - 5)$. Our intermediate step is to simplify $(-2x \cdot 4x) - (-2x \cdot 5)$. Just follow all of the rules we've been working with to get $-8x^2 - (-10x)$, for a final answer of $-8x^2 + 10x$.

$$-2x(4x - 5) = -8x^2 - (-10x) = -8x^2 + 10x$$

Distributing a variable with a negative coefficient

PRACTICE EXERCISES AND REVIEW

This chapter will likely be tricky for many students. It is important to not try to memorize the rules, but instead to just constantly remind yourself what exponents mean—namely repeated multiplication. Take some time to slowly and carefully review the chapter. Again, just focus on the concepts themselves.

For practice, try these exercises:

1) Simplify: $4x^3 \cdot 8x^5$
2) Simplify: $x^4 + x^9$
3) Simplify: $7x^8 \div 14x^2$
4) Simplify (with positive exponent): $16x^3 \div 4x^7$
5) Simplify: $(x^6)^8$
6) Evaluate: 7^0
7) Evaluate: $(-6)^1$
8) Simplify: $9xy^3z^4 \cdot 6x^3y^5z^9$
9) Distribute and simplify: $8x^3(5 + 7x^6)$
10) Distribute and simplify: $-4x^4(3x - 11)$

SO NOW WHAT?

Admittedly, this material can be confusing, and this is the point where many students get discouraged and frustrated. That is why it is so important to not move past any concept until you fully understand it, and can apply it quickly and accurately. Many problems that seem simple are actually dependent upon a dozen concepts being accurately combined and applied. Review this and all the previous chapters slowly and carefully before moving ahead to the next chapter.

CHAPTER FIVE

Solving and Graphing Algebraic Inequalities

WHAT IS AN INEQUALITY?

Informally, an **inequality** could be thought of as an equation which instead of having an equals sign has one of the following symbols: $<$, $>$, $\leq$, or $\geq$. Let's look at a simple example: $x + 2 > 7$. We need to determine what values of x will make this inequality true, namely that the left side

$$\begin{array}{r} x + 2 > 7 \\ -2 \quad -2 \\ \hline x \qquad > 5 \end{array}$$

is greater than 7. Let's assume that this problem had an equals sign instead of a greater-than sign. We could solve it algebraically or mentally to get $x = 5$. However, this problem involves a greater-than sign.

Think about why an answer of $x = 5$ will not work. When x equals 5, the left side is equal to 7, not greater than 7. To make the left side greater than 7, we need values of x that are greater than 5. Any value of x that is greater than 5 will satisfy the inequality. Our answer is $x > 5$.

Let's look at another example: $2x \leq 16$. That symbol is read as "less

$$\frac{2x}{2} \leq \frac{16}{2}$$

$$1x \leq 8$$

than or equal to." We're looking for all values of x which will make the left side either less than or equal to 16. Let's start by just looking at the "equal to" component. We know that if x is 8, the left side will equal 16. Therefore, 8 certainly must be part of our answer.

Now, what values of x will make the left side less than 16? Any value of x that is less than 8 will make the left side less than 16. For example, when x is 5, the left side is 10. What about 0? When x is 0, the left side equals 0 which is certainly less than 16. What about negative values of x? When x is -3, the left side equals -6 which is less than 16. Any value of x which is less than or equal to 8 will satisfy this inequality. Our answer is $x \le 8$.

"FLIPPING" THE INEQUALITY SYMBOL

It turns out that we can solve most algebraic inequalities in the exact same way that we solve algebraic equations, except that instead of using an equals sign in each step, we use the inequality symbol with which we started.

However, there is one very important rule to remember which exam-makers love to test students on. If we multiply or divide by a negative number at any time during the solving process, we must flip the inequality symbol so that it points in the opposite direction.

Let's look at an example, and see why that makes sense. Solve for x in the inequality -3x $\ge$ 12. If x equals -4, the left side will equal 12. We know that -4 will be part of the answer, but now do we need values that are less than -4, or greater than -4?

$$\frac{-3x}{-3} \ge \frac{12}{-3} \quad \text{Flip!}$$

$$1x \le -4$$

Let's spot-check some values. When x equals -2, the left side equals 6. That isn't good because we want to the left side to be greater than or equal to 12. Now let's try letting x equal -5. When x is -5, the left side is 15 which is good. If you test out other values including those less than and greater than -4, you will see that only the ones less than -4 satisfy the inequality. Our answer is $x \le -4$.

Since we divided by a negative value during the solving process, we needed to flip the direction of the inequality sign. We would do the same if we multiplied by a negative value at some point in the process.

Remember: When solving an inequality, if at any step you multiply or divide both sides by a negative value, you must flip the direction of the inequality symbol.

This may seem counterintuitive, so keep testing different values in the previous inequality until you are convinced that even though the problem involved "greater than or equal to," our answer involved "less than or equal to." Notice how important it is to be comfortable with signed number arithmetic. If you aren't, review the corresponding chapter in the first book.

$$\frac{3x}{3} \geq \frac{-12}{3}$$

No Flip!

$$x \geq -4$$

When solving an inequality such as $3x \geq -12$, we do not flip the inequality sign since we won't be multiplying or dividing both sides by a negative. The right side happens to have a negative, but we will be dividing both sides by 3 (a positive number) which does not alter the direction of the inequality sign.

GRAPHING INEQUALITIES ON A NUMBER LINE

A common task in algebra is to graph the result of a solved inequality on a number line. This is very easy if you just remember a few simple rules. Imagine that you solved an inequality, and ended up with a solution of $x > -2$. This result is illustrated graphically on the number line in the next example. We have a right-pointing arrow starting at -2 which shows that we are illustrating all values greater than -2. Even though the arrow stops, it is implied that it continues forever. We use an open circle at -2 to show that the value -2 is not itself included. This is because we are dealing with a "greater than" scenario as opposed to "greater than or equal to."

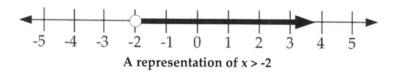

A representation of x > -2

Now imagine that you solved an inequality, and ended up with a solution of $x \leq 1$. This is illustrated below. We have a left-pointing arrow starting at 1 which implies all points less than 1. To show that the value 1 is itself included because of the "equal to" component of the "≤" symbol, we use a filled circle instead of an open one.

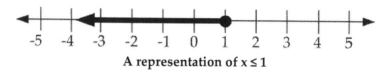

A representation of x ≤ 1

Let's look at one more common form of inequality: $-6 < (x - 2) \leq 1$. As strange as that may look, we are really just dealing with two inequalities in one. They can both be solved at the same time. We need to get x by itself in the middle which we would normally do by adding 2 to both sides. In this case, though, we actually have to add 2 to three different places—on the left, in the middle, and on the right. After doing so we have $-4 < x \leq 3$.

Think of that inequality as representing all values between -4 and 3. On the -4 side, we use an open circle. On the 3 side, we use a filled circle for reasons previously described. The inequality is depicted below.

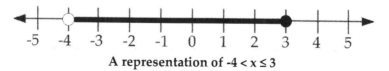

A representation of -4 < x ≤ 3

PRACTICE EXERCISES AND REVIEW

Remember: An inequality is solved in the same way that an equation is solved.

Remember: If any of the algebraic steps involve multiplying or dividing by a negative number, you must reverse the direction of the inequality. This is the only circumstance for which we do this.

Remember: When graphing an inequality on a number line, we use a filled circle to represent the condition of "less/greater than or equal to," and an open circle to represent the condition in which "equal to" is not included.

For practice, try these exercises:

1) Solve for x: $x - 4 \leq -11$
2) Solve for x: $-15x > 10$
3) Solve for x: $10x \geq -20$
4) Represent on a number line: $x \geq -2$
5) Represent on a number line: $-1 \leq x < 2$

SO NOW WHAT?

Before moving ahead, completely review all of the material that has been covered to this point. If anything is not completely clear, the next chapter will be extremely confusing. Also, review the basic math material covered in the first book to whatever extent is necessary.

CHAPTER SIX

FOILing and Factoring

SOME NEW TERMINOLOGY

In Chapter One, we informally defined an expression as a collection of terms connected by addition and/or subtraction. We also compared an expression to an equation by noting that an expression does not contain an equals sign whereas an equation does. For this chapter, we must define a few more terms that come up frequently in math. While you will not likely have to memorize the precise definitions of these terms, you must be familiar with them.

A **monomial** is an expression of one term. The prefix "mono-" means one. Some examples are 7, 3x, $19x^2y^3$, etc. Remember that a term is a collection of constants and variables connected via multiplication or division.

A **binomial** is an expression of two terms connected by addition or subtraction. The prefix "bi-" means two. Some examples are $(14 + 28x)$, $(x^2 - 16)$, $(8x^2 + 12x^3)$, etc. Parentheses were only added for clarity.

A **trinomial** is an expression of three terms connected by addition and/or subtraction. The prefix "tri-" means three. Examples are $(x^2 + 5x + 6)$ and $(3xy - 8x + 15y^3)$.

A **polynomial** is an expression of one or more terms, although it is most commonly used to describe an expression of at least three terms. The prefix "poly-" means many.

MULTIPLYING A BINOMIAL TIMES A BINOMIAL USING THE "FOIL" METHOD

A common task is to multiply two binomials together. An example is $(x + 2)(x + 3)$. Remember that the absence of an operation symbol between the binomials implies multiplication.

In Chapter Three, we worked with the distributive property as it applied to algebra. In particular, we distributed a monomial over a binomial. In this example, we will have to distribute a binomial over another binomial. It sounds much more complicated than it is. We just have to make sure that every term is multiplied by every other term, and that all combinations have been handled.

The acronym **FOIL** helps us keep track of those combinations. The F

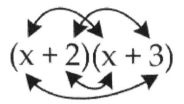

stands for "first." We'll start by multiplying the first terms of each binomial together. In this case we have $x \cdot x$ giving us x^2, and we will put that aside for the moment.

The O stands for "outer." Make sure you see how the x in the first binomial and the 3 in the second binomial could be thought of as the "outer" terms. When we multiply them together, we get $3x$ which we will put aside for now.

The I stands for "inner." Notice how the 2 in the first binomial and the x in the second binomial could be thought of as the "inner" terms. When we multiply them together, we get $2x$ which we will put aside for the moment.

The L stands for "last." The 2 and the 3 are the last terms in each binomial. When we multiply them together, we get 6, and we will put that aside for now.

Before moving on, take a moment to convince yourself that we handled every possible combination. There were two terms in the first binomial that had to be multiplied by the two terms in the second binomial. This makes the 4 combinations of FOIL. Note that if the first expression had 3 terms, and the second one had four terms, we would have $3 \cdot 4$ or 12 combinations to handle, but we wouldn't have any special acronym to help us.

Recall that we set aside four products during the FOIL process: x^2, 2x, 3x, and 6. The final step of the process is to add those products together, and combine like terms if applicable. We end up with $x^2 + 5x + 6$ which is the product of $(x + 2)(x + 3)$.

$$(x + 2)(x + 3) = x^2 + 5x + 6$$

Using the FOIL method to multiply two binomials

Remember: To multiply two binomials together, we use the FOIL method. This tells us to compute the product of the first terms, plus the product of the outer terms, plus the product of the inner terms, plus the product of the last terms. We compute the sum of the four products.

$$F.O.I.L = (Product\ of\ First) + (Product\ of\ Outer)$$
$$+ (Product\ of\ Inner) + (Product\ of\ Last)$$

Let's look at another example that involves some subtraction signs. Multiply: $(x + 4)(x - 7)$. The immediate question that most students ask is, "Does that dash mean subtraction, or does it mean that the 7 is negative?"

This was addressed in Chapter Three when we worked with the distributive property, but is worth revisiting since it is very confusing for most students. The main concept to understand is that $5 - 3$ is the same as $5 + (-3)$.

The first book outlined how we should convert signed number subtraction problems into addition problems that are easier to work with. Stated another way, we can always insert an implied addition sign in the middle of a subtraction problem, and use the minus sign as a negative sign for the value that it precedes.

With that said, for the sake of a FOIL problem, it is best to think of the subtraction signs as negative signs. For example, we will mentally convert the second binomial into $x + (-7)$, instead of $x - 7$.

Our original problem was $(x + 4)(x - 7)$. The product of the first terms is x^2. The product of the outer terms is $-7x$. The product of the inner terms is $4x$, and the product of the last terms is -28. Notice how we made our job a bit less confusing by simply treating the subtraction sign as a negative for the 7.

Our four products are x^2, $-7x$, $4x$, and -28. Remember that the second phase of the process is to add those products together. We get $x^2 + (-7x) + 4x + (-28)$. Parentheses were added for clarity. Combining like terms we get $x^2 + (-3x) + (-28)$. Recall that in situations like this, we can just omit the plus signs that precede the negatives, and treat the negatives as subtraction signs. We get a final answer of $x^2 - 3x - 28$.

$$(x + 4)(x - 7) = x^2 - 3x - 28$$

Multiplying two binomials which involve a negative

FACTORING A TRINOMIAL USING THE "REVERSE FOIL" METHOD

Another common task is to do what is called **factoring** a trinomial. This involves taking a trinomial such as $x^2 + 6x + 8$, and breaking it up into the product of two binomials. In a sense, we will be reversing the FOIL process which would have resulted in the given trinomial.

As expected, it sounds more confusing than it is. All we must do is some basic detective-style work using our math intuition, along with some "guessing and checking." Let's factor the above trinomial. We can start out by setting up our product of binomials as $(x \quad)(x \quad)$. We know that the only way we will end up with an x^2 is if the first terms are both x. Let's figure out how to end up with the +8 that we need. It will somehow come from multiplying the last two terms together. There are two ways that we can get 8, either $4 \cdot 2$ or $8 \cdot 1$.

For the next step, we just have to experiment to see which pair will work such that after FOILing we end up with the $6x$ that we also need to get. The $6x$ will be the result of adding the inner product plus the outer product. Let's try $(x + 1)(x + 8)$. If we use the FOIL method as described, we will end up with $x^2 + 9x + 8$. That doesn't work. If we try $(x + 2)(x + 4)$, we will end up with $x^2 + 6x + 8$ which is what we want.

$$x^2 + 6x + 8 = (x + 2)(x + 4)$$

Factoring a trinomial using "reverse FOIL"

In this particular problem, confirm that it would not have mattered if we switched the positions of the 2 and the 4. In later math, you will work with problems in which that is not the case.

Sometimes students think that teachers are withholding some secret method for factoring trinomials like the one we just worked with. They are not. The only way to solve these problems is using the thought process that we just used. With that said, as you gain more experience with these problems, the correct combination will "jump out" at you more quickly. You will just intuitively know that certain combinations won't work. Of course, as you improve your mental math ability, you will be able to quickly eliminate combinations that won't work.

Let's try another example: Factor $x^2 + x - 12$. As before, we can start out by setting up our product of binomials as $(x \quad)(x \quad)$. Next, let's think of how we can end up with negative 12. The possible combinations are:

$$6 \cdot (-2) \qquad 12 \cdot (-1) \qquad 4 \cdot (-3)$$
$$2 \cdot (-6) \qquad 1 \cdot (-12) \qquad 3 \cdot (-4)$$

We will need to test out each one, but again, with practice it gets faster and easier, and you will eventually just "see" the right choice.

Let's try FOILing $(x + 6)(x - 2)$. That gives us $x^2 + 4x - 12$ which doesn't work. Now let's try swapping the position of the minus sign so that we have $(x - 6)(x + 2)$. That gives us $x^2 - 4x - 12$ which also doesn't work. Now let's try FOILing $(x - 12)(x + 1)$. We get $x^2 - 11x - 12$. Swapping the position of the minus sign will make the $-11x$ positive instead of negative, but that won't help.

We need to get a middle term of $+1x$. With experience, you'll learn that picking last terms of 3 and 4 will accomplish that since they are just 1 apart. The question, then, is how to arrange the signs. Let's FOIL $(x - 4)(x + 3)$. We get $x^2 - x - 12$ which doesn't work. What we want is to put the plus sign with the 4 so that when we end up subtracting the $3x$, we will end up with positive $1x$. Confirm that the correct answer is $(x - 3)(x + 4)$.

$$x^2 + x - 12 = (x - 3)(x + 4)$$

Factoring a trinomial using "reverse FOIL"

FACTORING THE DIFFERENCE OF TWO SQUARES

A common task is to factor a binomial such as $x^2 - 16$ which is the difference of two squares. This looks different than the polynomials we've been working with because it only has two terms. If we wanted to make it look similar, we could actually put in a middle term, namely

$0x$ since that wouldn't change anything. Let's rewrite the binomial as $x^2 + 0x - 16$, and see how we can use the "reverse FOIL" method.

As before, we should start out by listing the various factor combinations that can make -16, but let's see if we can skip some steps. The only way we will be able to make the middle term of $0x$ is if we use a pair of factors that involve the same number. When we add the inner and outer products, we want the results to "cancel" each other out so we don't end up with a middle x term.

We need to make -16, and the only way we can do that is by multiplying a negative times a positive. Let's try FOILing $(x + 4)(x - 4)$, which is the correct solution. We get $x^2 - 4x + 4x - 16$. The two middle terms add to $0x$ which of course we can just omit. Confirm that using any other factor pairs such as $(x + 8)(x - 2)$ or $(x - 16)(x + 1)$ will not work since they will leave us with an unwanted middle x term.

$$x^2 + 0x - 16 = (x - 4)(x + 4)$$

Factoring the difference of two squares

It is important to understand that in our original binomial of $x^2 - 16$, we were dealing with the difference of two squares: x^2 is the square of x, and 16 is the square of 4. The technique that we used only works in such a situation. For example, we can factor $x^2 - 81$ as $(x - 9)(x + 9)$ using the same logic as the problem that we just solved.

In later math, you will work with factoring problems involving the difference of more complicated squares, but for now, the problems will all be similar to these.

Remember: The binomials we worked with involved the difference of two squares, and not the sum. They also involved perfect squares such as 36, 49, 64, etc.

"PULLING OUT" A COMMON FACTOR FROM A POLYNOMIAL

Many polynomials cannot be factored using the reverse-FOIL method that we worked with earlier, nor are they the difference of two perfect squares like those that we just worked with. An example would be $14x^5 + 21x^7$.

Whenever you see a polynomial in any context, it is usually best to first determine if there are any common factors among all of its terms. Stated another way, we will look for any constants or variables (perhaps including exponents) which divide into all of the terms, and we will factor them out. Sometimes students informally refer to this as "pulling" them out.

Let's revisit our example of $14x^5 + 21x^7$. Examining the coefficients of our variable terms, we can see that 7 is the GCF of 14 and 21. We can start by "pulling out" a 7, leaving us with $7(2x^5 + 3x^7)$. To verify that we didn't actually change the original expression, all we must do is use the distributive property to prove that we can get back to the original expression.

Let's look at what was left in parentheses. At this point, it is essential to fully understand everything presented in Chapter Four. Ignoring the coefficients, what is the GCF of x^5 and x^7? It is tempting to answer "1" if you are treating the 5 and 7 as though they are coefficients, but they are exponents, and therefore have a different significance.

Let's determine the highest power of x which divides into both x^5 and x^7. Does x itself divide into both? If we remember the rules for exponent division, we can see that $x^5 \div x = x^4$, and $x^7 \div x = x^6$. This shows that x does divide into both of those terms, which means it is a factor of both of those terms. This is quite tricky, so take some time to sit with this concept, and review the prerequisite material if necessary.

Let's try a higher power of x such as x^2. We can compute $x^5 \div x^2 = x^3$, and $x^7 \div x^2 = x^5$. This shows that x^2 is an even greater common factor of our two terms. Keep thinking about this concept until you see that the GCF of x^5 and x^7 is x^5. We can compute $x^5 \div x^5 = x^0 = 1$, and $x^7 \div x^5 = x^2$. If we try to use a higher power such as x^6, it won't divide into x^5. For example, $x^5 \div x^6 = x^{-1} = 1/x$ which is a fractional result.

This is a very tricky concept, so keep thinking about it. Make sure you understand the concept that the GCF of 37 and 41 is 1, but the GCF of x^{37} and x^{41} is x^{37}.

Getting back to our task of factoring $14x^5 + 21x^7$, we already pulled out a common factor of 7 from the coefficients, and we now see that we can pull out a common exponential term of x^5. After we pull out $7x^5$, what must go in parentheses so that if we distribute the $7x^5$, we will get back the original binomial? The answer is $7x^5(2 + 3x^2)$. Again, use the distributive property to verify that we can get back to the original binomial of $14x^5 + 21x^7$.

$$14x^5 + 21x^7 = 7(2x^5 + 3x^7) = 7x^5(2 + 3x^2)$$

Factoring a binomial by "pulling out" common factors.
With practice, the intermediate step can be skipped.

Now let's try factoring $6x^4 - 30x^{10}$. Notice we can "pull out" a common factor of 6 from the coefficients. Next, following what we just learned, confirm we can factor out x^4. That is the GCF of x^4 and x^{10}. We factored out $6x^4$, and are essentially dealing with this: $6x^4(? - ?)$. What must go in parentheses so that after distributing the $6x^4$, we end up with our original binomial? We must think "in reverse."

What we must do is divide each of the original terms by the $6x^4$ that we factored out. When we divide $6x^4$ by $6x^4$, we actually get 1. Give that some thought. When we divide $30x^{10}$ by $6x^4$ and follow our rules for exponent division, we get $5x^6$. This means that the factored version of the original binomial is $6x^4(1 - 5x^6)$. Again, this can be quite tricky. Think about it for as long as necessary, and review the previous chapters as needed.

$$6x^4 - 30x^{10} = 6(x^4 - 5x^{10}) = 6x^4(1 - 5x^6)$$

Factoring a binomial by "pulling out" common factors.
With practice, the intermediate step can be skipped.

FACTORING THE SQUARE OF A BINOMIAL

Let's factor $x^2 + 10x + 25$ using the "reverse FOIL" method. As before, we will set up our framework as $(x \quad)(x \quad)$. We know the last two terms will have to be either 25 and 1, or 5 and 5. We can experiment with both combinations, but with experience, you will recognize that it is the pair of fives we need. We also must use pluses instead of minuses, otherwise we won't end up with a positive middle term.

The correct factorization is $(x + 5)(x + 5)$. Notice that we are really dealing with the same binomial multiplied by itself, which is synonymous with squaring. This means that we can rewrite our answer as $(x + 5)^2$, which is how such an answer will often be represented. Conversely, if we started out with $(x + 5)^2$, and were asked to multiply or "expand" $(x + 5)^2$, we should start out by writing it as $(x + 5)(x + 5)$, and then use FOIL as we've learned.

$$x^2 + 10x + 25 = (x + 5)(x + 5) = (x + 5)^2$$

A factoring which results in the square of a binomial

Let's try another example without going into as much detail: Factor $x^2 - 14x + 49$. We'll start with our framework of $(x \quad)(x \quad)$. The last two terms will have to both involve 7, but in this case, they will have to both be negative. When FOILed, this will give us the +49 third term, as well as the -14x middle term. Confirm that the correct factorization is $(x - 7)(x - 7)$, which can be written more concisely as $(x - 7)^2$.

$$x^2 - 14x + 49 = (x - 7)(x - 7) = (x - 7)^2$$

A factoring which results in the square of a binomial

In later math, you may work with formulas for problems that follow these patterns. For now, just be sure that you are comfortable with the idea that some polynomials will factor into the square of a binomial, and that the binomial itself may involve a sum or a difference.

SIMPLIFYING ALGEBRAIC FRACTIONS

Students are justified in asking the question, "Why do we need to factor something when it looks just as complicated after factoring as before?" As usual, one answer is that this skill comes into play in later math, but of course, many students are not required to reach that level, nor are they interested in doing so.

It is worth examining a more advanced problem that involves factoring. A common problem in later math is to simplify a division of polynomials such as the ones in this example:

$$\frac{x^2 + 4x - 21}{x^2 - 8x + 15}$$

A division of polynomials that requires factoring in order to simplify

The "plain English" explanation for how to proceed is that we must factor both the numerator and before we can do any "canceling." Remember that the whole point of factoring is to rewrite an expression as a product of terms or expressions. Once we are dealing with nothing but products, we can "cancel" common factors.

Stated another way, as the problem stands now, we are not allowed to just "cancel" the matching x^2 terms, nor are we allowed to "pull out" a common factor of 4 from the +4 and -8 that we see. We can't do this because the numerator and denominator each involve sums and differences, and not products. Confirm that after factoring the numerator and denominator using the "reverse FOIL" method, we have:

$$\frac{(x - 3)(x + 7)}{(x - 5)(x - 3)} = \frac{(x + 7)}{(x - 5)}$$

The factored polynomials (which are now products) can be simplified by "canceling" matching terms

Now that we have nothing but products in the numerator and denominator, we can "cancel" matching terms or expressions. In this case, we can "cancel" the $(x - 3)$ in the numerator and denominator, remembering that each really becomes a 1 which we don't have to write. We are left with $(x + 7) / (x - 5)$, which is how we must leave the answer. We cannot cancel the x terms, nor could we cancel anything else even if it seemed possible since the numerator and denominator involve a sum and a difference, and not products.

This topic will be practiced extensively in later math. For now, remember that a common reason for factoring expressions is so that we can simplify them. Another reason you will study later is that factored expressions are easy to work with. For example, in the expression $(x - 4)(x + 6) = 0$, what values of x will make the equation true? We don't really need to do any math. If $x = 4$, the expression in the first set of parentheses becomes 0, making the entire left side 0. If $x = -6$, the same is the case for the second set. This means that the solutions are $x = \{-4, 6\}$ (meaning either-or). Again, you will work with this concept in later math. For now, just notice that we wouldn't be able to solve this problem visually if it started out as the unfactored $x^2 + 2x - 24 = 0$.

PRACTICE EXERCISES AND REVIEW

There is little to review in this chapter that does not involve rewriting what has already been written. Take the time to review this very important chapter slowly and carefully. One of the most important and confusing points is reviewed below.

Remember: The GCF of 29 and 31 is 1. This is because if we try to divide those two numbers by any number greater than 1, we will get a fractional result. This means that there is no common factor to "pull out" of $29x + 31y$ other than the "moot point" 1.

Remember: In contrast to the above example, the GCF of x^{29} and x^{31} is x^{29}. Review the rules of exponent division to confirm that if we were to divide those two values by x^{10}, for example, we would get x^{19} and x^{21}, respectively. Neither of those are fractional values which proves that x^{10} is a common factor (but not the greatest one). The GCF of x^{29} and x^{31} is x^{29} because after dividing each of those two values by x^{29}, we get 1 and x^2, respectively. Neither of those are fractional values which proves that x^{29} is a common factor, and in fact the greatest one.

Remember: A higher value such as x^{30} is not a factor of x^{29}. By the rules of exponent division, $x^{29} \div x^{30}$ is x^{-1} which is equal to $1/x$ which is a fractional value. Any value above x^{30} will also result in a fractional value.

For practice, try these exercises:

1) Multiply with FOIL: $(x - 6)(x + 2)$
2) Multiply with FOIL: $(x - 8)(x - 5)$
3) Factor: $x^2 + 5x + 6$
4) Factor: $x^2 - 7x + 12$

5) Factor: $x^2 - 64$

6) Factor: $x^2 - 225$

7) Factor: $10x^8 - 15x^{12}$

8) Factor: $4x^6 + 8x^{17}$

9) Factor and simplify: $x^2 + 10x + 25$

10) Factor and simplify: $x^2 - 14x + 49$

SO NOW WHAT?

The next chapter begins the unit on basic geometry. Since geometry problems often involve working with basic algebra, this would be a good time to review all of the material that has been covered so far. It is also absolutely essential that you feel fully comfortable with topics in basic math, especially signed number arithmetic. Review the material in the first book as needed.

CHAPTER SEVEN

Introducing Geometry: Lines and Angles

WHAT IS GEOMETRY?

Geometry is a vast field of study in math, but unless you are planning to study advanced math, you will only be responsible for scratching the surface of it. In a general sense, geometry is the study of points, lines, shapes, and solids, and how they relate to one another.

SOME BASIC DEFINITIONS AND CONCEPTS

The basic building block in geometry is a **point**. A point is informally defined as a location in space. We could draw a dot on a piece of paper, chalkboard, or some two-dimensional surface to represent a point. If we had some way of drawing a dot in midair, that would represent a point in a three-dimensional space such as a room.

If we take a point, and extend it in opposite directions, we form a line. A **line** is informally defined as a set of points arranged in a straight pattern, extending infinitely in both directions. We use an arrow on each end to signify this. If a line has a specific stopping point on both sides, we refer to it as a **line segment**. If a line stops on one end, but extends infinitely in the other direction, we refer to it as a **ray**.

Line Segment Line Ray

We can form two-dimensional shapes or figures by connecting lines in different ways. Some examples are squares, rectangles, triangles, circles, or odd-looking shapes that don't have a particular name. We say that these shapes have two dimensions because we can measure them in two distinct directions which for now we can refer to as their length and width.

The instructions for a problem will sometimes say that several points or a two-dimensional figure lies in a plane. Informally, we define a **plane** as a flat surface. The instructions are effectively telling us that the points or figure could be thought of as drawn on a flat sheet of paper or on the chalkboard.

We can form three-dimensional solids by taking a two-dimensional shape, and extending it into the third dimension. One way to visualize this is to imagine taking a shape on a piece of paper or chalkboard and extending it "upwards" or "outwards" off the surface.

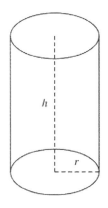

See if you can visualize that if you extend a circle in this way, it becomes a cylinder, and if you extend a rectangle in this way, it becomes a rectangular solid, which could informally be thought of as a box. Even though we typically use the word solid to refer to three-dimensional structures, they are not necessarily "filled in." In fact, a common task in geometry is to determine exactly how much space there is inside a three-dimensional structure.

If we agree that a shape such as a square has two dimensions, and a solid such as a cube has three dimensions, how many dimensions does a line have? A line is defined as having just one dimension. We could think of it as length. We define a line as being infinitely thin. When we draw a line on paper or on the chalkboard, it must have a tiny bit of thickness just so we can see it, but that doesn't count as a dimension.

If a line has one dimension, how many dimensions would a point be? A point is defined as having zero dimensions. We may draw a big fat point on the chalkboard so that it will be easy to see, but we don't actually measure its thickness in any way. It is just an infinitely small representation of a location in space.

USING LINES TO FORM ANGLES

When two rays extend outward from a shared point (the **vertex**), an **angle** is formed. Angles are also formed when two lines intersect (i.e., cross each other) at a point. We can informally define angle in this way.

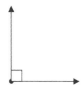

One of the most common types of angles is a **right angle**. Later, we will use the word perpendicular as the defining characteristic of a right angle, but for now, you can informally think of it as anything that looks like the angle at left, or any rotation of such.

Right angles are common in everyday life such as the corners of pieces of paper or non-rounded tables. You may know from experience that a right angle is defined as having 90°. Let's see where that comes from.

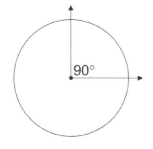

We define a circle as having 360°. If we draw an angle so that the vertex is at the center of the circle, we can use basic math to determine what percentage of the circle is represented by the angle. In the picture at left, note that a right angle represents exactly ¼ of a circle. We can compute 360 ÷ 4 to get 90 which is where 90° comes from.

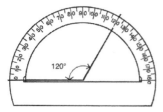

We can use a **protractor** to determine the exact measurement of angles, but you will likely only be responsible for knowing just a few basic types. If we take two rays that share a common vertex, and arrange them on top of each other, a 0° angle is formed. As we widen the distance between the two rays, the angle measurement starts to increase.

For example, the picture at left shows a 45° angle. Informally, we can

say that it is halfway between a 90° right angle and a 0° angle in which the two rays are the same. We use the term **acute** to define any angle that is greater than 0° and less than 90°.

What would happen if we started with a 90° right angle, and increased the separation of the two rays by rotating one of them further around a circle? Eventually, the rays would point in exact opposite directions. In terms of a circle, we would have half of it. We could compute 360 ÷ 2 to get 180 which is the degree measurement of such an angle. A 180° angle is called a **straight angle**.

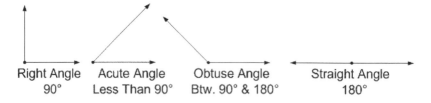

| Right Angle | Acute Angle | Obtuse Angle | Straight Angle |
| 90° | Less Than 90° | Btw. 90° & 180° | 180° |

If an angle is greater than a 90° right angle, but less than a 180° straight angle, we refer to it as an **obtuse angle**. An example is shown above. You may also encounter the term **reflex angle**. A reflex angle is an angle that is greater than a 180° straight angle, but less than a 360° angle that would be a full rotation around a circle.

COMPLEMENTARY AND SUPPLEMENTARY ANGLES

If the measures of two angles add to 90° (i.e., a right angle), we say that the angles are **complementary**. A problem might state that two angles are complementary, and one of them measures 38°. The problem would ask for the measure of the other angle which is easily computed as 90 minus 38 for an answer of 52°.

If the measures of two angles add to 180° (i.e., a straight angle), we say that the angles are **supplementary**. A problem might state that two angles are supplementary, and one of them measures 131°. The problem would ask for the measure of the other angle which is easily computed as 180 – 131 for an answer of 49°.

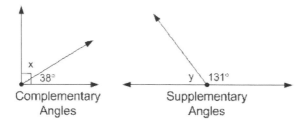

Complementary Angles   Supplementary Angles

It may be helpful to remember that the C of "complementary" alphabetically precedes the S of "supplementary," just as 90° numerically precedes 180°.

ANGLE PROBLEMS INVOLVING ALGEBRA

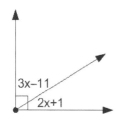

If you memorize the definition of these terms, and if you have mastered the basic algebra techniques presented in this book, you should have no trouble handling problems of these types that lead to algebra. Look at diagram at left. If we were asked to solve for x, we would set up our equation as $(3x–11) + (2x+1) = 90$, and solve it algebraically. Confirm that the answer is $x = 20$. The answer can be checked by substituting it into the expression which represents each angle, and confirming that the two angles, in this case 41° and 49°, add up to 90°. Such a problem could have also been presented without a diagram by stating that the two given expressions represent the measures of two complementary angles.

When solving problems like this, only answer the specific question asked. If asked to solve for a variable, just solve for it. Don't substitute its value into the expressions to determine the angle measurements. If asked to determine the measures of each angle, don't answer with the value of the variable that you determined. Just substitute it in each expression. If asked to determine the measure of the smaller or larger angle, answer with that and only that. You won't get extra credit for answering with more information—your answer will be wrong.

PARALLEL AND PERPENDICULAR LINES

Informally, we say that two lines are **parallel** if they will never meet. We can say that they have the same slant or slope. Look at the picture below for an example.

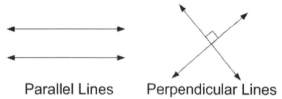

Parallel Lines   Perpendicular Lines

We say that two lines are **perpendicular** if they intersect each other to form right angles. Look at picture above. It is not mandatory that perpendicular lines be horizontal and vertical. Any rotation of such would still be sufficient as long as the lines met at right angles.

WORKING WITH INTERSECTING LINES

When two lines cross each other at a point, we say that the two lines **intersect.** It is the same concept as roads that form an intersection.

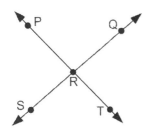

Notice that when two lines intersect, four angles are formed. In the diagram at left, point R is the **vertex** of each of the four angles. However, we cannot speak of "angle R "(symbolically: $\angle R$) without providing more information since it would be unclear as to which of the four angles it represents.

In the diagram above, we represent the angle at the top as either $\angle PRQ$ or $\angle QRP$. Notice how the vertex is named in between the two endpoints. Each of the other three angles can be named in a similar fashion, each with vertex R in the middle.

PROBLEMS INVOLVING PARALLEL LINES INTERSECTED BY A TRANSVERSAL

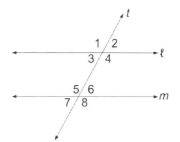

In the diagram at left, we see two lines that are assumed parallel. They are intersected by a third line called a **transversal**. Notice how eight angles are formed. Problems involving a diagram like this are common and simple.

Sometimes, we are asked to indicate the special name assigned to a particular pair of angles. Angles 1 and 4 are called **vertical** angles or **opposite** angles. They will always be equal in measure regardless of the angle at which the traversal intersects the parallel lines.

Angles 2 and 3 are also vertical or opposite angles. They are equal in measure, but not equal to angles 1 and 4. Notice that there are also two pairs of vertical angles involving the lower parallel line.

Angles 1 and 5 are called **corresponding** angles, and have equal measure. Notice how they are in corresponding relative positions. For practice, note the three other pairs of corresponding angles in the diagram. Remember that whenever two angles lie along a straight line (i.e., straight angle), they are called supplementary. The sum of any two such angles is 180°. In the previous diagram, angles 1 and 2 are supplementary. Angles 1 and 3 are also supplementary. They lie along a straight angle which happens to run diagonally. For practice, find the other pairs of supplementary angles in the diagram.

Angles 3 and 6 are called **alternate interior** angles because they lie on alternate sides of the transversal, and are inside the two parallel lines. They are equal in measure to each other. Angles 4 and 5 are also alternate interior angles. They are equal in measure to each other, but are not equal in measure to angles 3 and 6.

Angles 1 and 8 are called **alternate exterior** angles because they lie on alternate sides of the transversal, and are outside the two parallel lines. They are equal in measure to each other. Angles 2 and 7 are also alternate interior angles. They are equal in measure to each other, but are not equal in measure to angles 1 and 8.

Once you have memorized the names of these special angle relationships (assuming you are required to), the main concept to understand is that "small" angles equal "small" angles, and "big" angles equal "big" angles.

You also must understand that the sum of a "small" angle and a "big" angle will be 180°. All this is true regardless of the angle at which the transversal intersects the two parallel lines. It is also worth noting that all these angle properties are contingent upon the two main lines being parallel to each other, but problems will always state or imply such.

You may encounter problems on this topic that involve algebra. See the related section earlier in the chapter for the general guidelines to follow. The only challenge is to properly set up your equation. The problem will involve either two angles that are equal in measure, or two angles that sum to 180°. In the case of the former, just set the two expressions equal to each other. For the latter, just add the two equations together, and set the result equal to 180°. Once that is done, just solve the equation algebraically as we've learned, and be careful to only answer the specific question asked.

PRACTICE EXERCISES AND REVIEW

This chapter included many definitions that you should review and memorize. Some important concepts are reviewed next.

Remember: Complementary angles sum to 90°, and supplementary angles sum to 180°. These definitions must be memorized since they won't be provided.

Remember: Algebra problems involving geometry will almost always be a matter of either setting two expressions equal to teach other, or adding two expressions and setting the sum equal to 90 or 180. Read the problem or study any accompanying diagram carefully to find out. With that, just solve the equation algebraically using the techniques previously taught.

Remember: Be careful to answer only the specific question asked in a problem. Don't answer with any extra information, and don't answer a completely different question. Read questions slowly and carefully.

Remember: Problems involving parallel lines and a transversal typically come down to the concepts of "big" versus "small" angles. All big angles are equal to all other big angles, and all small angles are equal to all small angles. The sum of a small angle and a big angle is 180.

<u>For practice, try these exercises:</u>

1) State the degree measurement/range of these angles:
   a) right; b) straight; c) acute angle; d) obtuse
2) Two angles are complementary. One measures 39 degrees. What is the degree measure of the other?
3) Two supplementary angles are represented by the expressions $(3x - 28)$ and $(7x + 58)$. Find the measure of the larger angle.
4) Two vertical angles are represented by the expressions $(2x + 4)$ and $(7x + 14)$. Find the value of x.

SO NOW WHAT?

The next chapter offers an introduction to the geometry topics of area, perimeter, and volume. These concepts rely heavily on topics in basic math, so review the material in first book as needed.

CHAPTER EIGHT

Area, Perimeter, and Volume

COMPUTING THE AREA OF FIGURES

In math, we define **area** as the space inside a two-dimensional (i.e., flat) figure. Imagine that you want to install wall-to-wall carpeting or tiling in a room of your house. We are dealing with the floor which is a effectively a flat shape. The pricing of the material and service will likely be based upon the area of the room in question (i.e., how much space needs to be filled).

When we state the area of a figure, we never just use a number by itself. If someone says, "The area is 49," what does that mean? Does it mean inches, feet, yards, miles, etc.? We can't be certain. It is also incorrect to include just the unit, since the units in question measure length, and not actual space. This can be tricky, but make sure you understand that if someone says "49 yards," they are referring to an actual distance as measured with a straight line. When we speak about area (i.e., the space inside of a shape), we must always speak in terms of square units. This will make more sense in a moment.

When we measure the area inside a shape, what we are really deter-mining is how many unit squares can fit inside that shape. This is the case regardless of what shape we are dealing with. The shape need not be a square or a rectangle, although those are certainly easier for visualizing the concept of area.

Imagine a closet whose floor measures 7 feet by 3 feet. To compute the area, we must determine how many squares can fit inside it. Since the dimensions have been measured in terms of feet, we must determine how many square feet can fill the space of the floor.

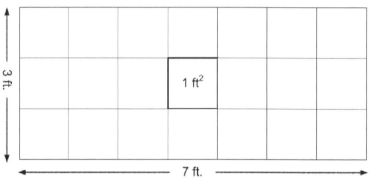

Computing the area of a room in terms of square feet

A square foot is defined as a square that measures 1 foot on each side. This is a common size and shape of floor tiles. Note that if we were dealing with a shape drawn on paper, we might work with square centimeters or square inches depending on its stated dimensions.

Returning to our example, notice that we can fit 21 square feet inside the given shape. This was effectively the result of multiplying 3 times 7, which makes sense. It is important to understand that if you were to state the area as 21, it would be marked wrong. It would also be marked wrong if you stated the answer as 21 feet. That answer implies a length (i.e., a distance), and not an area (i.e., a space). The correct answer is "21 square feet" (21 sq. ft.) which is more commonly represented as 21 ft^2.

Some students get confused, and think we're dealing with 21^2 which is 441. That is not how it works. Our answer of 21 ft^2 means that we can fit 21 squares inside our shape with each one measuring 1 ft. by 1 ft.

Note that if an area problem provides dimensions to work with, but does not state the units involved (e.g., inches, feet, yards, etc.), we must answer the question in terms of square units, abbreviated $units^2$.

THE AREA OF RECTANGLES AND SQUARES

Area problems typically do not involve a diagram of boxes like the one we worked with, nor will you be expected to draw such boxes. We solve area problems using special formulas for common shapes. In some cases, we might have to take a complex shape, and draw lines to break it up into common shapes.

We just learned the formula for computing the area of a rectangle. We multiply its two dimensions which are commonly referred to as length and width. Again, we must remember to write our answer in "units squared" form, substituting the name of the involved unit if stated.

$$Area_{(rectangle)} = Length \times Width$$

A square is just a special type of rectangle in which all four sides are equal. It makes sense that its area formula will be similar, and it is. In a sense, we must multiply the length times the width, but in a square, those two dimensions are equal. Each could be thought of as a side. The formula is $A = Side \times Side$ which can be shortened to $A = Side^2$.

$$Area_{(square)} = Side^2$$

WHY DO WE ALWAYS WORK WITH SQUARE UNITS?

There is one final point of confusion to clarify before discussing the area formulas for common shapes. As mentioned, regardless of what shape we are dealing with, we always answer area questions in terms of square units. If we are computing the area of a triangle, we do not answer in terms of "triangle units," nor do we compute the area of a circle in terms of "circle units." We always deal with square units.

For shapes that are not squares or rectangles, instead of using a diagram, we must use special formulas to compute the area since it is difficult to actually count squares inside the shape. Even if we could, some of the squares would have to be cut so they could properly fit.

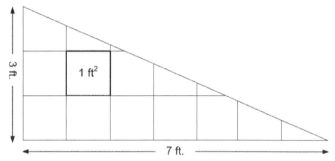

The area of a triangle is still measured in square units

The triangle above is exactly half of the rectangle we worked with. We can fit some full squares inside, but we must use partial squares along the diagonal edge.

If the area of the rectangle was 21 ft^2, and this triangle is half that in size, it makes sense that its area would be half as well, namely 10.5 ft^2. We can see six full squares in the triangle. If we were to arrange the other segments like a jigsaw puzzle, we would make another 4.5 squares.

COMPUTING THE AREA OF A TRIANGLE

From the sample triangle drawing, we can deduce that the formula for the area of a triangle is half that of a rectangle. This is actually the case regardless of the type of triangle that we are working with. It may be tempting to conclude that the formula is $A = \frac{1}{2} \times Length \times Width,$ but that is not exactly the case.

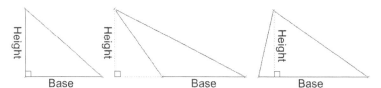

Notice how the height is measured in each model of triangle.

Look at the three triangle models above. When we compute the area of a triangle, we must speak in terms of base and height. The rule is the height is always measured from the highest point of the triangle "straight down" (i.e., perpendicular) to the base. In the first triangle, the height is measured along one of the sides since that side was already perpendicular to the base.

In the second triangle, the highest point is not directly over the base. This means we must extend the base using an imaginary line, and drop our height using an imaginary line drawn straight down from the highest point to the extended base. When computing the area of such a triangle, we will only use the actual base. We do not include the extension.

In the third triangle, the highest point is over the base. We just drop an imaginary line straight down to the base, and use that as the height.

With those terms and concepts established, we can now say the area of a triangle is ½ times the base times the height. It is important to understand that we never measure the height along a diagonal (slanted) side of a triangle. The height is always measured perpendicular to the base. It will be either inside or outside the triangle, or along the side of a right triangle. If the lengths of unrelated sides of a triangle are provided, it is just decoy data unless part of a more advanced problem.

Remember: A triangle's height is measured from its highest point straight down (perpendicular) to the base.

$$Area_{(triangle)} = \frac{1}{2} \times Base \times Height$$

AREA OF A PARALLELOGRAM

In non-math terms, a **parallelogram** could be thought of as a slanted

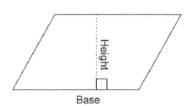

rectangle. It has two pairs of parallel sides as shown. A rectangle is just a special parallelogram in which one pair of parallel sides happens to be perpendicular to the other pair.

Since a parallelogram is related to a rectangle, it makes sense that its area formula would be similar. The only difference is that instead of computing length times width, we must compute base times height, with the height defined in the same way that it was for triangles, namely perpendicular to the base.

$$Area_{(paralellogram)} = Base \times Height$$

Remember: If you are given the lengths of the slanted sides, it is decoy data intended to trick you unless that data is necessary to compute the actual height. Since you won't face such problems until later math, it is very likely that you will be given the actual perpendicular height which is what you should use in the formula.

AREA OF A TRAPEZOID

A **trapezoid** is a **quadrilateral** (four-sided figure) which has one pair of parallel sides and one pair of non-parallel sides as shown at right.

Again, the formula involves the height which is measured from the top base straight down (perpendicular) to the lower base.

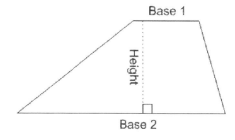

The area formula below looks more complicated than it is. The formula is just the height times the average of the two bases. This makes sense since a trapezoid is effectively a rectangle with one of its sides shortened, and the other two sides slanted to keep the shape enclosed.

$$Area_{(trapezoid)} = \frac{(Base_1 + Base_2)}{2} \times Height$$

The same advice about decoy data applies to trapezoids. If we are told the lengths of each of the two bases, as well as the perpendicular height, then any information about the lengths of the slated sides is extraneous.

Remember: Area is always stated in terms of square units regardless of the shape.

COMPUTING THE PERIMETER OF A FIGURE

In math, we define **perimeter** as the distance around a figure. We just add up the lengths of all the sides, and include the given unit in our answer. We don't work with square units since there is no multiplying involved, and we are not measuring the space inside the figure. We are just measuring the distance around it.

It is important to not fall for trick questions. For example, you will sometimes need to apply your knowledge of shapes to fill in missing data. If we are told that two sides of a parallelogram measure 3 inches and 5 inches, we can deduce that the other two sides are 3 and 5 inches, respectively, for a perimeter of 16 inches.

Look at the triangle below. The instructions would likely state that it

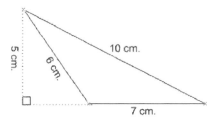

was "not drawn to scale." This means we should just use the numbers given, and not worry about whether they "look right." Confirm that the area of this triangle is 17.5 cm^2, and the perimeter is 23 cm. To compute the area, we ignored all data besides the 5 and 7, and to compute the perimeter, we ignored the height of 5 since it is not part of the border of the actual shape.

TERMINOLOGY RELATED TO CIRCLES

It is important to learn and memorize some basic terminology related to circles. In the circle below, line segment AD (symbolically: $\overline{AD}$) is called a **radius**. It is a line that starts at the center, and extends to a

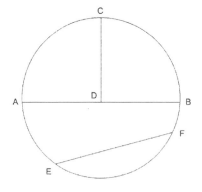

point on the edge of the circle. $\overline{DC}$ and $\overline{DB}$ are also radii (plural). An infinite number of radii can be drawn starting at center D, and extending to any point on the circle itself.

$\overline{ADB}$ is called a **diameter**. It is a line that starts along the circle itself, and extends straight through the center to end on the opposite side. An infinite number of diameters can be drawn by starting at any point on the circle and drawing a line straight through the center as described. A diameter need not be vertical or

102

horizontal. A diameter is twice the length of a radius, and conversely, a radius is half the length of a diameter.

$\overline{EF}$ is called a **chord**. It starts at some point on the edge of the circle and extends to another edge, but it does not cross through the center as does a diameter.

Although we use the word perimeter to refer to the distance around a multisided shape, we use the word **circumference** to mean the distance around the edge of a circle. Think of it as the perimeter of a circle.

WHAT EXACTLY IS PI (π)?

The Greek letter **pi (π)** occurs very frequently in geometry. Before we discuss how it is used, it is important to learn what it actually is, and where it comes from.

Imagine drawing a circle of a convenient size. If you wanted to measure its circumference, you could get a reasonably precise value by wrapping string around the border, and then measuring the length of the string used. You would probably measure it in inches. You could then measure the diameter using a basic ruler, again in inches.

π If you were to take the circumference, and divide it by the diameter, the units of inches would "cancel," and you would be left with a value of approximately 3.14. We call that value pi(π). We say that π is the ratio of a circle's circumference to its diameter. We would get the same value of about 3.14 regardless of how big or how small we drew the circle.

You will work with π much more if you study advanced math. For now, be sure to understand that 3.14 is only an approximate value that we commonly use for π. Its actual value is a non-repeating decimal (known as an irrational number) which continues infinitely without a pattern (3.14159265358979323846264338327950288...) Your teacher or textbook may instruct you to use 22/7 as an approximate value for π, but again, it is just an approximation. Some exam questions test to see if you understand that 3.14 and 22/7 are not exact values of π.

COMPUTING THE AREA OF A CIRCLE

Now that we learned the general concept of π, let's see how it is used when computing the area of a circle. The formula for the area of a circle is $A = \pi r^2$. We must take the radius, square it, and then multiply that result by pi using whatever approximate value we are told to use (likely 3.14). Again, always state the answer in terms of square units.

$$Area_{(circle)} = \pi \times radius^2$$

Let's try a problem that involves an extra step: Find the area of a circle whose diameter is 9 cm. Round the answer to the nearest tenth. First, note that we were given the diameter, but the formula calls for the radius. The radius is half the diameter, which in this case is 4.5 cm. We square that to get 20.25, and then multiply that by 3.14, our approximate value of pi. We get 63.6 (rounded), and we must write our answer as 63.6 cm^2.

COMPUTING THE CIRCUMFERENCE OF A CIRCLE

Recall that the circumference of a circle is effectively its perimeter (i.e., a distance or length). The formula for the circumference of a circle is $C = 2\pi r$ **or** $C = \pi d$. Confirm that twice the radius equals the diameter, which is why both formulas are equivalent. Choose whichever formula is more convenient based on the data provided.

$$Circumference\ of\ Circle = 2\pi r = \pi d$$

Let's try an example: A circle has a radius of 6 feet. What is its circumference rounded to the nearest tenth? Let's use the first version of the formula. After substituting, we must compute 2 times pi times 6, giving us 37.7 (rounded). Since we are dealing with length and not area, we will write our answer as 37.7 feet.

Remember: The formula for the area of a circle involves r^2 which makes sense area involves multiplying two dimensions. The formula for circumference is derived from the fact that C ÷ d = π. Since circumference is just a measure of length, there is no squaring involved.

COMPUTING THE VOLUME OF SOLIDS

So far, we've been working with two-dimensional (i.e., flat) shapes. We learned how to compute their perimeter and their area. We will now work with the three-dimensional counterparts to those shapes. They are often referred to as solids, but we will learn how to compute the space inside them, known as the **volume**.

Imagine taking a square or rectangle, and extending it outward from a piece of paper or chalkboard. You would end up with a three-dimensional object that is commonly thought of as a box. In math, we refer to such an object as a **rectangular solid** or **rectangular prism**.

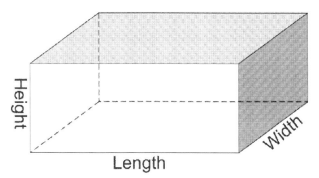

A three-dimensional rectangular solid (e.g., a box)

Recall that to compute the area of a rectangle, we just multiplied its two dimensions (usually called length and width). It follows that to compute the volume of its three-dimensional counterpart, we would multiply all three dimensions involved, with the third dimension usually referred to as the depth or height.

$$Volume_{(rectangular\ solid\ or\ rectangular\ prism)} = Length \times Width \times Depth$$

WORKING WITH CUBIC UNITS

Recall that the area of a shape is based on how many unit squares can fit inside it. We always use squares regardless of the actual shape which is why our answer must include $units^2$, or whatever specific unit is involved.

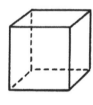 If we extend this concept to three dimensions, it makes sense that instead of filling our object with flat squares, we will fill it with three-dimensional unit cubes. This is hard to draw on a two-dimensional piece of paper, and it may be hard to visualize.

Imagine an empty box that measures 12 inches by 8 inches by 5 inches. If we wanted to fill in the box using unit cubes whose edges measure one inch, we could fit 12 of them along one dimension, 8 along another, and 5 along the third. To compute the number of 1-inch cubes which will fit inside the box, we just have to multiply the three dimensions, giving us 480 $inches^3$. We read that answer as "480 cubic inches," or "480 inches cubed."

Don't get confused and think that we are dealing with 480^3 which is over 110 million. Our answer means that we can fit 480 cubes inside the box, with each cube measuring 1" by 1" by 1". Just as we always used squares when computing area regardless of shape, we will always use cubes when computing volume.

THE VOLUME OF A RIGHT CIRCULAR CYLINDER

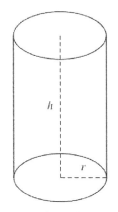

At left is a **right circular cylinder**. It is formed by taking a circle and extending it upwards or outwards from a flat surface into the third dimension. The word "right" is included in its name to emphasize that the circle is extended perpendicular to the surface on which it was drawn (i.e., it isn't leaning).

It is difficult to draw and visualize how many unit cubes can fit inside a rounded cylinder. Obviously, the cubes along the edges would have to be cut and rounded. Instead, let's just deduce the formula based on what we know. We know that the formula for the area of a circle is πr^2. That takes care of two of the three dimensions of the cylinder. The third dimension was created by just adding height or depth to the circle. To compute the volume of a cylinder, all we must do is compute the area of the circle upon which it is based, and then multiply it times the third dimension. As always, we will express our answer in terms of cubic units.

$$Volume_{(right\ circular\ cylinder)} = \pi r^2 \times height$$

Let's try an example: "A right circular cylinder has a diameter of 7 cm, and a height of 10 cm. What is its volume rounded to the nearest tenth? (Use 3.14 for π)." First, we need to compute the radius which is 3.5 cm. Square that, multiply it by 3.14, then multiply it by 10 to get 384.7 (rounded) which we must write as 384.7 cm^3.

THE VOLUME OF A TRIANGULAR PRISM

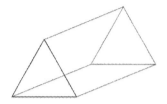

A **triangular prism** is a formed by taking a triangle and extending it into the third dimension. While it may be tricky to visualize, we can apply our knowledge of how volume works to deduce its volume formula. Just as we did with a cylinder, first we will find the area of the triangle upon which the solid is based. That takes care of two of the three dimensions. We then multiply that value times the third dimension, usually called the height or depth.

There is one tricky aspect to this that is worth noting. A problem will likely refer to the height of the triangle upon which the prism is based, since height is a component of the area of a triangle. The problem may also use the term height as opposed to depth to describe the extension of the triangle into the third dimension. To make matters worse, both of those heights may be abbreviated in a diagram using the letter "h." With all that said, understand that the height of the triangular face of the prism has nothing at all to do with the height (i.e., extension of that triangle into the third dimension). To summarize, just compute the area of the triangular face using the appropriate formula, and then multiply it by the third dimension value given.

$$Volume_{(triangular\ prism)} = \frac{1}{2}bh \times depth$$

THE VOLUME OF OTHER SOLIDS

Until later math, it is not likely that you will be asked to compute the volume of solids which were not covered so far. If you are, it is very likely that the volume formula for the given solid will be provided for you, and you will just need to substitute given values. Of course, if you are expected to memorize any formulas, you should do so.

Three common solids that were not yet covered are the cone, square pyramid, and sphere. The formula for the volume of a cone is $V = \frac{1}{3}\pi r^2 h$. This implies that the volume of a cone is ⅓ of a cylinder which makes sense since a cone is essentially a tapered cylinder.

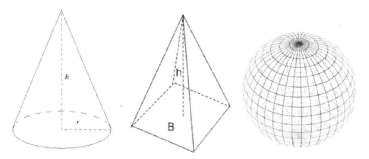

From left to right: A cone, square pyramid, and sphere

The formula for the volume of a square pyramid is $V = \frac{1}{3}Bh$. A square pyramid has a square base, and four triangular sides which come together to form a point at a the top. In this formula, capital B is being used to represent the area of the square base of the pyramid. That area itself is computed as $side^2$ as we learned. We then multiply that by the height of the pyramid, and multiply that result by ⅓. Notice that the relationship between a square pyramid and a rectangular solid is analogous to that of a cone and a cylinder.

The formula for the volume of a sphere is $V = \frac{4}{3}\pi r^3$. It makes sense that the formula would involve πr^3, since the area of a two-dimensional circle is πr^2. The 4/3 part just needs to be memorized.

THE EFFECT OF INCREASING DIMENSIONS ON THE AREA (OR VOLUME) OF FIGURES (OR SOLIDS)

A common exam question states that the dimensions of a shape or solid have been increased in a certain way, and asks us to determine what effect such a change has on the area or volume of the shape or solid.

Think about the area formulas that we studied. Regardless of the shape, they all involved multiplying two dimensions together. They might be called length and width, or base and height, or in the case of a circle, simply the radius times itself.

What would happen if we doubled one of the dimensions but not the other? For example, what happens to the area of a 3 × 5 rectangle if we make it 6 × 5? The area doubles from 15 to 30. Think about why this makes sense, not just as far as the arithmetic, but visually as well. If you double the length of your garden, but leave the width as is, you'll have twice as much room to plant.

What would happen to the area if you doubled both of the dimensions, taking the 3 × 5 rectangle and making it 6 × 10? The area quadruples from 15 to 60. Again, think about why this makes sense. We have a doubling times a doubling, resulting in four times the area.

What would happen to the area if you doubled the length, but tripled the width? Since the area formula is length times width, we must take into account that the doubling is being multiplied by the tripling, result in the new area being six times as large as the original.

Be careful about problems of this type that involve circles. If you were to triple the radius, the area of the new circle would be nine times as big as the original. This is because the formula requires us to square the radius, which means we must square the tripling, resulting in a new area that is 3^2 or 9 times as big.

This pattern holds for all shapes, and for any multiplicative changes (e.g., doubling, tripling, etc.) made to one or both dimensions. Just multiply any stated changes to determine the overall change.

As expected, all of this extends to three-dimensional solids. For example, confirm that if the length, width, and depth of a rectangular solid were each tripled, it would increase the volume of the solid by 27 times (i.e., triple times triple times triple).

In the case of a sphere, any change to the radius must be cubed, as per the volume formula. Confirm that multiplying a sphere's radius by 4 (i.e., quadrupling it) results in a new sphere that is 4^3 or 64 times as big as the original.

IMPORTANT POINTS TO REMEMBER

Area is always expressed in square units, and volume is always expressed in cubic units. You will lose points unnecessarily if you do not express your answers as such.

The area of a triangle is ½bh where h is defined as the distance from the highest point on the triangle straight down to the base (or extension).

The perimeter of a figure is the distance around its border. Remember to include the units.

The diameter of the circle is the distance from edge to edge though the middle, and the radius is half that. The area formula of a circle is πr^2, and the circumference formula is $2\pi r$ or πd.

We usually use 3.14 as an approximate value of π, but π is actually non-repeating and non-terminating.

Volume formulas are derived from computing the area of a solid's base, and multiplying it by the third dimension, usually denoted height or depth.

If a figure's or solid's dimensions change (e.g., double, triple, etc.), multiply the changes to determine the net effect on its area or volume.

<u>For practice, try these exercises:</u>

1) Find the area of a 9 cm by 7 cm rectangle.
2) Find the area of a square of side 25 yards.
3) Find the area of a right triangle with legs of 5 and 12 inches, and a hypotenuse of 13 inches.
4) Find the area of a parallelogram with a base of 8 cm, slanted sides of 5 cm each, and a height of 4 cm.
5) Find the area of a trapezoid with bases of 9 ft. and 10 ft., height 3 ft., and slanted sides of unknown length.
6) True/False: $\pi = 3.14$
7) True/False: π is approximately 22/7
8) Find the area of a circle of diameter 16 cm.
9) Find the circumference of a circle of radius 5 ft.
10) Find the volume of a rectangular solid of 3 by 8.6 by 4.7 feet.
11) Find the volume of a cube of edge 3 cm.
12) Find the volume of a cylinder with circular base of area 21 cm^2 and height of 8.5 cm.
13) Find the volume of a triangular prism of depth 20 in. whose triangular base has a height of 4 in. and a base of 6 in.
14) A circle's radius is tripled. How many times larger is the area of the resulting circle?
15) A rectangular solid has both its length and depth tripled, and its width quadrupled. How many times larger is the area of the resulting solid?

SO NOW WHAT?

The next chapter covers the basics of the Pythagorean Theorem and triangles, both of which are popular topics on geometry exams. Before proceeding, review the concepts of proportion and squaring. Also, review the algebra section on solving more complicated equations.

CHAPTER NINE

The Pythagorean Theorem and Triangles

THE SUM OF INTERNAL ANGLES IN A TRIANGLE

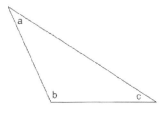

In any triangle, the sum of the three interior angles is always 180°. This is true regardless of the type of triangle. Many exam questions test to see if you know this. The question will provide the measures of two of the three angles, either in writing or with a picture, and you must compute the third angle. This is easily done by subtracting from 180.

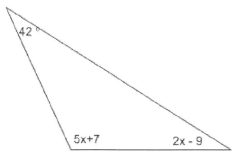

Many problems on this topic involve basic algebra. If one or more of the angles involve a variable, just set up an equation in which the sum of the three angles is set equal to 180. As always, read the question carefully. If you are asked to compute the value of x, answer that and nothing more. You may also be asked to compute the measures of each angle which is easily done once you have algebraically solved for x. In the diagram above, we can solve for x using the equation $5x + 7 + 2x - 9 + 42 = 180$.

Some questions ask only for the measure of the smallest or the largest angle. If you list more than one angle or the wrong angle, the question will be marked wrong.

COMMON TYPES OF TRIANGLES

There are four types of triangles whose names and properties you must memorize. In the picture below, the first triangle is an **equilateral triangle**. It has three equal sides and three equal angles. This is implied by each side having a single tick-mark, and each angle having a single arc. Since the sum of the internal angles in any triangle is 180°, and all three angles are equal, we can deduce that each angle is 60°.

The second triangle is an **isosceles triangle**. It has two equal sides, and two equal angles as implied by the tick-marks. A typical algebra problem may present an expression for each of the two "base angles" of an isosceles triangle. Knowing the properties of an isosceles triangle, it is easy to solve for x by setting up an equation with those two expressions set equal to each other.

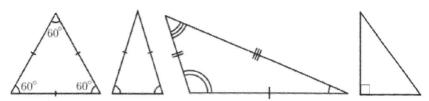

From L to R: Equilateral, isosceles, scalene, right triangle

The third triangle above is a **scalene triangle**. It has three unequal sides and three unequal angles. We worked with a scalene triangle in the previous section. Algebra problems involving scalene triangles typically require adding the expressions that represent each angle, and setting the sum equal to 180.

The fourth triangle is a **right triangle**. The only implied information is that it contains a 90° (right) angle. Algebra problems involving right triangles typically require adding the expressions representing the other two angles, and setting the sum equal to 90.

A **right isosceles triangle** is formed by combining the attributes of right and isosceles triangles. It has a right angle, two equal sides, and two equal angles, leading to the picture at right.

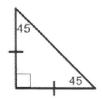

SIMILAR AND CONGRUENT TRIANGLES

Some word problems involve the term **congruent**. In math, the term "congruent" just means "identical to." If two figures are congruent, they are the exact same size and the exact same shape. One of the two figures might be rotated, but it is still effectively an exact photocopy.

In English, the word "similar" can mean different things to different people. In math, "similar" means "proportional to." Stated another way, two figures are **similar** if they have the same shape, but are different sizes. For example, if we take a particular triangle, and double the length of each side, the resulting triangle will be the same shape, but proportionally larger.

We use a tilde (~) to symbolically mean "is similar to." For example, in the following diagram (not drawn to scale), a problem may stipulate that $\triangle ABC \sim \triangle DEF$, and ask us to determine the values of the missing sides, x and y.

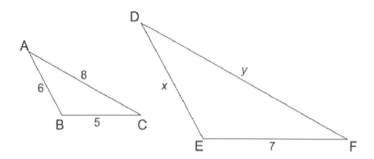

$$\frac{5}{7} = \frac{6}{x}$$

To do this, we must set up a proportion, and solve it as we learned in Chapter Three. One way to solve for x is to note that 5 is to 7 just as 6 is to x (base is to base just as left side is to left side). Our proportion is $5/7 = 6/x$. Cross multiply as we learned to get $5x = 42$, which means $x = 8.4$. For practice, confirm that we also could have adopted the mindset that 5 is to 6 just as 7 is to x (base is to left side just as base is to left side). Our proportion would be $5/6 = 7/x$. This leads to the same equation and same answer.

To solve for y with the first method, confirm that we would set up the proportion $5/7 = 8/y$. Cross multiply to get $5y = 56$, meaning $y = 11.2$.

A common word problem on this topic involves two adjacent objects such as a person, tree, or flagpole each casting a shadow. Since the sun is in the same position relative to both, we are really dealing with a problem of similar right triangles as illustrated in the diagram for the following example:

A 6-foot tall man casts a 5-foot shadow. What is the length of an adjacent girl's shadow if she is 2'6" tall, and her shadow is measured at the same time of day?

116

As with the previous problem, there are two equivalent ways of setting up our proportion:

$$\frac{man's\ height}{man's\ shadow} = \frac{girl's\ height}{girl's\ shadow} \quad \textbf{or} \quad \frac{man's\ height}{girl's\ height} = \frac{man's\ shadow}{girl's\ shadow}$$

Let's use the first model, and set up our proportion as $6/5 = 3.5/x$. Notice how we converted the half foot of height into a decimal, and how we assigned a variable to the component of the diagram that we wanted to compute. Cross multiplying yields the equation $6x = 12.5$, which means $x = 2.9$ feet (rounded to the nearest tenth).

THE PYTHAGOREAN THEOREM

A very common topic in geometry involves what is known as the **Pythagorean Theorem**. Before discussing it, some basic terminology must be defined. In a right triangle, we define the longest side as the **hypotenuse**. It will always be the side opposite the right angle since the right angle will always be the biggest angle in triangle. We call each of the other two sides a **leg** of the triangle.

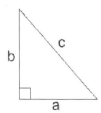

The Pythagorean Theorem states that in any *right* triangle, if you take the square of the length one leg, and add it to the square of the other leg, the sum will equal the square of the hypotenuse. In the following diagram, sides a and b are legs, and c is the hypotenuse. Symbolically, the Pythagorean Theorem states that $a^2 + b^2 = c^2$.

Problems involving the Pythagorean Theorem typically give us the values of two of the three sides of a right triangle, and ask us to compute the length of the unknown side. Sometimes the unknown side is a leg, and sometimes it is the hypotenuse. The information may be provided in the form of a picture or a word problem.

$$a^2 + b^2 = c^2$$

The Pythagorean Theorem

Let's try a problem: A right triangle has legs of 3 cm and 4 cm. What is the length of the hypotenuse? After substituting the values into the theorem, our equation is $3^2 + 4^2 = c^2$. After simplifying, we have $25 = c^2$. Recall that "square rooting" is the inverse of squaring, so take the square root of each side to get c = 5 cm.

You may realize that -5 is also a solution to the above equation since $(-5)^2$ also equals +25. However, in geometry we deal with lengths, and lengths can never be negative so we just discard that solution.

Always check your answers for reasonableness. If a right triangle has legs of 3 and 4 cm, the hypotenuse is certainly longer than 4 cm. It can't be much longer, though, because then the legs would make an angle wider than 90°. An answer such as 25 would certainly be wrong.

Let's try another problem: A right triangle has a leg of 5 ft., and a hypotenuse of 13 ft. What is the length of the second leg? After substituting the values into the theorem, our equation is $5^2 + b^2 = 13^2$. In this case, our unknown value is a leg—not the hypotenuse. After simplifying, we have $25 + b^2 = 169$. Subtract 25 from each side to get $b^2 = 144$. Take the square root of each side to get $b = 12$ ft.

This is the pattern that we follow for any problem involving a right triangle with an unknown value. Note that very often the numbers will not work out as well as they did in these problems. Let's try this problem: A right triangle has a leg of 4 ft., and a hypotenuse of 8 ft. What is the length of the second leg? After substituting the values into the theorem and evaluating the squares, our equation is $16 + b^2 = 64$. Subtract 16 from each side to get $b^2 = 48$ (which is not a perfect square). Take the square root of each side to get $b = \sqrt{48}$ ft.

In later math, you'll learn how to simplify a square root such as $\sqrt{48}$. For now, just follow the instructions in the problem. They may tell you to leave the answer in square root form as is, or to use your calculator to evaluate the square root, rounding the answer to a given place. For example, $\sqrt{48}$ rounded to the nearest hundredth is 6.93.

COMMON PYTHAGOREAN TRIPLES

Most Pythagorean Theorem problems in later math don't have integer answers because the missing side is typically not a perfect square. Until then, though, many of the Pythagorean Theorem problems you will face will involve what are called **Pythagorean Triples**.

Some right triangles have sides whose lengths just work out "nicely." For example, we already worked with a triangle whose sides measured 3, 4, and 5 units, and with a triangle whose sides measured 5, 12, and 13 units. These sets of lengths are known as Pythagorean Triples.

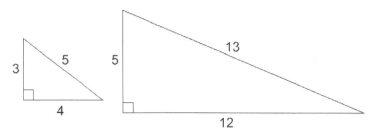

Common Pythagorean Triples

Since it is common for Pythagorean Theorem problems to involve triangles with sides represented by one of these sets, it's best to memorize them so there won't be any need to actually use the Theorem.

Many problems on this topic involve a multiple of one of these sets of Triples. For example, if a right triangle has sides of length 6 and 10 units, the missing side must be 8 units. Such a triangle will be similar to a 3-4-5 triangle—each side is twice as long. As another example, if a right triangle has sides of length 36 and 39 units, the missing side must be 15 units. Such a triangle will be similar to a 5-12-13 triangle—each side is three times as long.

PRACTICE EXERCISES AND REVIEW

Remember: The sum of the internal angles in any triangle is 180°. If the angles are represented by expressions containing a variable, set up an equation in which the expressions are summed and set equal to 180.

Remember: Only answer the specific question asked in a problem (e.g., the value of x, the measure of the smallest or largest angle, etc.)

Remember: The Pythagorean Theorem is only applicable to right triangles. It states that the sum of the squares of the legs is equal to the square of the hypotenuse.

Remember: To solve problems involving similar triangles (whether presented in word or diagram format), just set up a proportion, and set the cross products equal.

Remember: The most common Pythagorean Triples are {3, 4, 5} and {5, 12, 13}, and their multiples.

For practice, try these exercises:

1) A triangle has angles of 38.1 and 102.5 degrees. What is the degree measurement of the third angle?
2) The angles of a triangle are represented by the expressions $(2x - 11)$, $(4x + 5)$, and $(7x - 9)$. Find the degree measure of the smallest angle.

3) Describe the characteristics of an isosceles triangle.

4) An isosceles triangle has base angles represented by the expressions $(8x + 6)$ and $(2x + 8)$. Solve for x.

5) A 5 ft. tall pole casts a 7 ft. shadow. How tall is a nearby tree if its shadow measures 13 ft.?

6) A right triangle has legs of 10 and 24 ft. Find the hypotenuse w/o using the Pythagorean Theorem.

7) A right triangle has a leg of 8 ft., and a hypotenuse of 13 ft. What is the length of the other leg?

SO NOW WHAT?

The final chapter introduces linear equations and plotting points on a coordinate plane (graph). Before proceeding, be sure that you have no difficulty with signed number arithmetic such as $(-7) - (-3)$, and $3 - 5$. Also, review the section on solving more complicated equations.

CHAPTER TEN

Linear Equations and the Coordinate Plane

THE CARTESIAN COORDINATE PLANE

In geometry, we often work with points that are plotted on what is called a **Cartesian coordinate plane**, informally known as a **graph**. We typically use graph paper with small squares to draw one.

The first step in drawing a coordinate plane is to draw a horizontal line known as the **x-axis**, and a vertical line known as the **y-axis**. For our purposes, it should be sufficient if each line is about 15 boxes in length, and if the two lines intersect near their midpoints.

Look at the following coordinate plane. Each axis is really just a number line including a negative and positive side, with 0 in the middle. Notice how the right side of the x-axis is labeled "x," and the negative side is labeled "-x." Notice how the y-axis is really the same number line, but rotated so that it runs vertically. It is labeled "y" at the top, and "-y" at the bottom. Get in the habit of labeling your axes in this way to avoid unnecessarily losing points on exams.

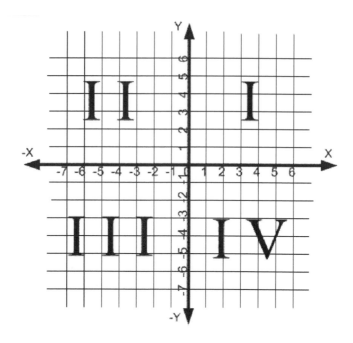

Notice how each intersecting line on each axis has been labeled just as we would do with the tick-marks on a number line. We always label the lines themselves, and not the spaces between them. This step is called **scaling** the graph. We are showing the viewer that each line represents one unit. In news articles, you may come across graphs in which each line represents a larger quantity. Be sure to always scale your graphs as shown.

A Cartesian coordinate plane is divided into four **quadrants**. The upper right hand quadrant is known as the first quadrant, sometimes abbreviated QI. In the upper left, we have the second quadrant, or QII. The lower left quadrant is the third quadrant or QIII, and in the lower right, we have the fourth quadrant or QIV.

A typical exercise is to **plot** a point at a given place on the graph, or to determine the **coordinates** (i.e., the address) of a plotted point.

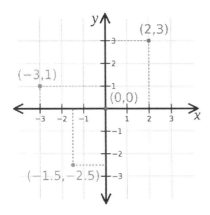

Look at the coordinate plane at left. The point at which the two axes intersect is known as the **origin**. It has coordinates of (0, 0). In our examples, we will think of the origin as our "home" from which we will walk a number of "blocks" horizontally and then vertically to get to other points on the graph.

Look at the point labeled (2, 3). To determine the coordinates of a point, we start at the origin, our "home." We first determine how many "blocks" left or right (or West or East) we must walk to be aligned with our destination. We always must walk left or right before walking up or down. In this case we must walk two blocks right. Since we walked to positive 2 along the x-axis, we say that the x-coordinate of the destination point is 2.

Now that we have completed the left/right component of our walk, we must walk up or down (or North or South) to get to the actual destination point. In this case, we must walk three blocks up. Since we are effectively aligned with positive 3 on the y-axis, we say that the y-coordinate of our destination point is 3.

The coordinates of a point are given as an **ordered pair** in the form (x, y). The x-coordinate is always listed before the y-coordinate, just as x comes before y alphabetically. The coordinates of our destination point are (2, 3). This tells the reader to start at the origin (0, 0), and move two lines to the right, and three lines up. An ordered pair is just walking directions to a location in a town whose laws require us to always walk West/East before turning and walking North/South.

Let's look at some other points on that graph. The point (-3, 1) has those coordinates because to reach it, we must first walk three blocks left (in the negative direction on the x-axis), and then 1 block up. To reach the point (-1.5, -2.5), we must first walk 1½ blocks left, and then 2½ blocks down, in the negative direction on the y-axis.

Remember: To find the coordinates of any point, we always start our "walk" at the origin (0, 0). The coordinates of a point are never given relative to another point that we may have already plotted.

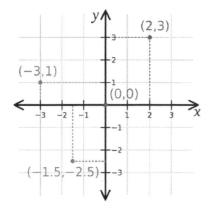

Another way of thinking about all of this is that we can determine the coordinates of a point by observing what value on each axis the point is in line with. The dotted lines in the above graph help us to see that, although it is not necessary to include them on any graph that you sketch. The most important thing to remember is that we always list the x-coordinate first, followed by the y-coordinate. A coordinate can be negative, positive, or 0 which we will work with next.

A point can lie on the x or y-axis itself. In fact, we've been working with the point (0, 0) which is simultaneously on both axes. To determine the coordinates of a point on an axis, we follow the same rules as we did for any other point. The only difference is that to get to a given point, we will not have to travel both vertically and horizontally. Some examples will make this clear.

Look at the uppermost point on the coordinate plane below. To determine its coordinates, we will start at the origin as always. Remember that our first step is to always walk left/right (West/East). In this case, we don't have to walk at all. We are already horizontally lined up with the point. Since we didn't walk left or right along the x-axis, the x-coordinate of our destination point is 0. Now we must determine how far up/down (North/South) we must walk to reach our point. We must walk 5 blocks up, which means the y-coordinate of our point is 5. The coordinates of the point are (0, 5).

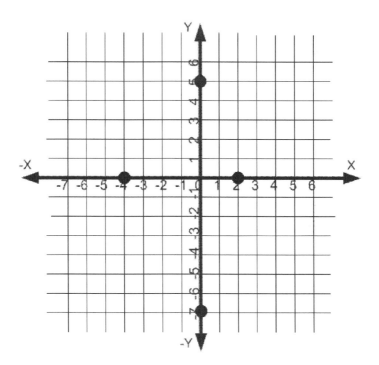

Notice that for points on an axis itself, we cannot use the "dotted line" method of determining a point's coordinates which was demonstrated in an earlier example. That is why it is preferable to just ask yourself, "How far left/right, and then how far up/down?" keeping in mind that 0 is an acceptable answer to either or both of those questions.

Look at the rightmost point on the previous graph. To determine its coordinates, we will start at the origin as always. Remember that our first step is to always walk left/right (West/East). In this case, if we walk two blocks right along the x-axis, we are at our destination point. There is no need to walk up or down. Since we didn't, the y-coordinate of our destination point is 0. The coordinates of the point are (2, 0).

Using the same procedure, confirm that the coordinates of the leftmost point are (-4, 0), and the lowermost point is (0, -7). It is easy to get confused when working with points that lie along on one of the axes. Just remember all of the rules that we follow. First, we compute the left/right x-coordinate, then we compute the up/down y-coordinate. One or both of those values may be 0. We write our ordered pair in the form (x, y). The x-value must always come first.

COMPUTING THE SLOPE (*m*) OF A LINE

In everyday English, the term "slope" can mean different things to different people, but in math, it has a precise definition, and is computed by using a specific formula.

To compute the **slope** of a line, we must first determine the coordinates of any two points along the line. A problem will either give you the coordinates of the two points to use, or the points will be marked on the line. Again, any two points along the line can be used, but you should avoid working with points whose coordinates are not obvious because they do not fall precisely on an intersection of the grid. Look at the dotted line on the graph at right. Two points with coordinates (2, 4) and (-1, -5) are marked.

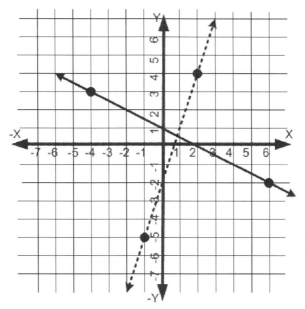

There are five different versions of the formula for slope (always denoted *m*), all of which mean the same thing.

$$m = \frac{change\ in\ y}{change\ in\ x} = \frac{y_A - y_B}{x_A - x_B} = \frac{vert.\ change}{horiz.\ change} = \frac{rise}{run} = \frac{\Delta y}{\Delta x}$$

Five equivalent representations of the formula for slope (*m*)

The first representation states that we can determine the slope of a line by computing the difference between the y-coordinate of each chosen point, and dividing it by the difference between the x-coordinate of each point.

To compute the difference between two values, we just subtract. This is illustrated in the second representation of the slope formula. To compute the slope of a line, choose one point on the line and call it Point A. Some books and teachers may call it Point 1. Choose another point on the line and call it Point B (or Point 2). It doesn't matter which point you choose to be Point A (or Point 1) as long as you are consistent throughout your calculations.

All we do is take the y-coordinate of the first point, and subtract the y-coordinate of the second point. We then similarly compute the difference between the x-values of the points, being careful to not reverse the order of what we are calling each point. We take the change in y that we computed, and divide it by the change in x.

The last three versions of the formula are just other ways of saying the exact same thing. The change in y could be thought of as the vertical change, and the change in x could be thought of as the horizontal change. We sometimes refer to the change in y as the "rise," and the change in x as the "run." However, these terms can be confusing when we are working with a line which does not slope up and to the right. The final representation uses the Greek letter delta (Δ) which means "change in."

The previous coordinate plane has been repeated at right. Let's compute the slope of the dotted line using the second version of the formula. Let's call the upper point A, and the lower point B, but again, we could have done the opposite as long as we're consistent.

The coordinates of Points A and B are (2, 4) and (-1, -5), respectively. The change in y is computed as 4 – (-5) which is 9. The change in x is computed as 2 – (-1) which is 3. Compute 9/3 to get 3, our slope.

Note that we would get the same slope even if we swapped the order of what we called Point A and Point B. If we did, we would compute the change in y as (-5) – 4 which is -9, and the change in x as (-1) – 2 which is -3. Computing -9/-3 still gives us the same slope of +3.

Understand that you will not be able to compute the answers to slope problems if you do not feel fully comfortable with signed number arithmetic. Take time to review that topic in the first book.

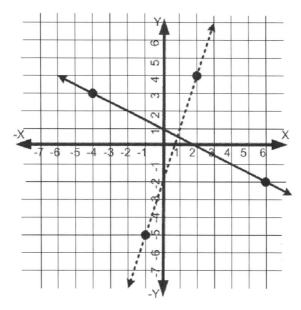

Now let's compute the slope of the solid line. The coordinates of the upper point are (-4, 3). Let's call it Point A. The coordinates of the lower point are (6, -2) which we'll call Point B. We'll compute the change in y as 3 − (-2) which is 5. We'll compute the change in x as -4 − 6 which is -10. Take the change in y and divide it by the change in x to get a slope of 5/-10 which simplifies to -½. For practice, confirm that we would get the exact same slope if we swapped which points we called A and B.

FOUR "DIRECTIONS" OF SLOPE

We've seen a line with a positive slope, and one with a negative slope. A line can also have a slope of 0, or a line can have what we call "no slope" or "undefined slope," neither of which are the same as a line with a slope of 0.

Any line that slants up and to the right has a **positive slope**. Depending on which point you call Point A and which you call Point B, the slope formula will result in a positive divided by a positive, or a negative divided by a negative. In either case, the result is positive.

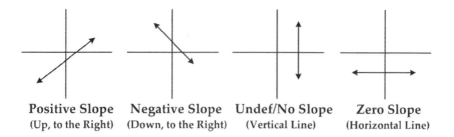

Positive Slope	**Negative Slope**	**Undef/No Slope**	**Zero Slope**
(Up, to the Right)	(Down, to the Right)	(Vertical Line)	(Horizontal Line)

Any line that slants down and to the right has a **negative slope**. Depending on which point you call Point A and which you call Point B, the slope formula will result in a positive divided by a negative, or a negative divided by a positive. Either way, the result is negative.

A vertical line has what we call an **undefined slope** or **no slope**. This is *not* the same as a zero slope. If we compare any two points on a vertical line, there will be a vertical change, but no horizontal change. Stated another way, there will be a change in y, but not in x. Since the slope formula has the change in x in the denominator, and the change in x is effectively 0, we have a "divide by 0" situation which is undefined. This is why all vertical lines have an "undefined slope" or "no slope." A way to remember this is that you cannot drive your car down the side of a building. You would just "free fall."

A horizontal line has **zero slope**. This is *not* the same as an undefined slope or no slope. If we compare any two points on a horizontal line, there will be a horizontal change, but no vertical change. Stated another way, there will be a change in x, but not in y. Since the slope formula has the change in y in the numerator, and the change in y is effectively 0, the fraction we end up with is of the form $0/n$. We know that 0 over anything is just 0. This is why all horizontal lines have a zero slope. A way to remember this is that you can certainly drive your car on a perfectly level road. It just isn't sloped up or down.

THE SLOPE AS AN INDICATION OF "STEEPNESS"

A line's slope also gives us an idea of its steepness, regardless of whether it is positive or negative. In other words, the sign of a slope (i.e., positive or negative) tells us if the line slants up or down, but the slope's magnitude (i.e., the value when ignoring the sign) indicates the line's steepness.

Think about what it means if a line has a slope of 1. Since it is positive, the line will slant up and to the right. Since the slope formula is (change in y) ÷ (change in x), the only way we can get a slope of 1 is if the values used in the slope formula are of the form *n/n*. Stated another way, as the line makes its way up and to the right, the vertical change must always equal the horizontal change. This would result in a line that makes a 45° angle as can be seen in the diagram below.

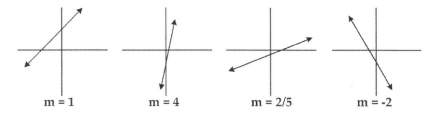

m = 1   m = 4   m = 2/5   m = -2

A line's steepness can be inferred from its slope's magnitude

Now think about what it means if a line has a slope of 4. The ratio of the change in y to the change in x is 4:1. Stated another way, as we trace the line up and to the right, we will move 4 units up for every one unit that we move to the right. Since 4 is greater than 1, it makes sense that a line of slope 4 will be steeper than a line of slope 1.

Look at the line of slope 2/5 in the diagram. A slope can have a fractional value. The ratio of the change in y to the change in x is 2:5. Stated another way, as we trace the line up and to the right, we will move 2 units up for every 5 units that we move to the right. We can say that the "run" (rightward movement) of the line exceeds the "rise" of the line. In everyday terms, this results in a line with a shallow slant, at least more so than a line with a slope of 1. This makes sense since 2/5 is less than 1.

Look at the line of slope -2. Since it is negative, the line will slant down and to the right. The ratio of the change in y to the change in x is 2:1, ignoring the negative. As we trace the line down and to the right, we will move 2 units down for every 1 unit that we move to the right. Compare the relative steepness of the various lines in the diagram, disregarding the fact that the negative slope line slants in a different direction than the others.

INTRODUCING THE Y-INTERCEPT (*b*)

We've seen that a line's slope paints a picture of its direction and steepness. The **y-intercept** of a line helps to describe its actual position on the coordinate plane. The y-intercept tells us where the line crosses the y-axis. All lines other than vertical ones will eventually cross the y-axis somewhere if they are extended far enough.

Look at the first line in the diagram at right. It crosses the y-axis at 0, so we say that its y-intercept (always denoted *b*) is 0. Note that the slope of the line does not play a part in making this determination.

Look at the second line, again disregarding its slope. We can see that it crosses the y-axis at some point which is negative on the y-axis. Since the coordinate axes have not been scaled, we can't state a precise value for *b*, but it will definitely be less than 0.

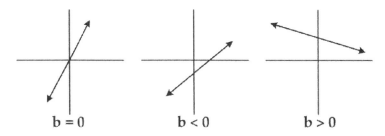

$$b = 0 \qquad b < 0 \qquad b > 0$$

From L to R, lines with 0, negative, and positive y-intercepts

Look at the third line, again disregarding its slope. We can see that it crosses the y-axis at some point which is positive on the y-axis. Since the coordinate axes have not been scaled, we can't state a precise value for b, but it will definitely be greater than 0. A line's slope and y-intercept are used in computing what is known as the equation of the line, discussed next.

EQUATION OF A LINE IN SLOPE-INTERCEPT FORM

We only need two pieces of information to describe any line on the coordinate plane—its slope and its y-intercept. The slope of the line (m) tells us the direction in which it slants and its steepness. The y-intercept (b) identifies a point on the line, namely the point (0, b). We know that the x-coordinate at the y-intercept is 0 since any point along the y-axis has an x-coordinate of 0.

The slope and y-intercept are used in what is called the equation of a line in **slope-intercept** form. The general pattern of the equation is **y = mx + b**. In this equation, the x and y are variables. In fact, a typical exercise is to choose any value of x that you wish, and then algebraically solve for y, or vice-versa. The m and b in the equation are typically provided. Some examples will make all of this clear.

Let's look at the equation y = 3x + 2. If we match up this equation with

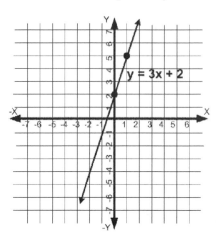

the general equation of y = mx + b, we can see that the slope (m) is 3, and the y-intercept (b) is 2. This tells us that the point (0, 2) will definitely be on the line. The fact that the slope is positive tells us that the line will slant up and to the right. The slope value of 3 (effectively 3/1) tells us that the ratio of the vertical increase to the horizontal increase is 3:1.

Stated another way, to get from one point on the line to the next, we go up 3, and right 1. All we must do is plot one more point on the line besides (0, 2). Then we can draw a line extending through those points.

Let's look at the equation y = -2x – 3. Matching this up to the general

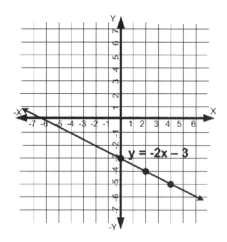

equation of y = mx + b, we can see that the slope is -2. That means that the line will slant down and to the right. The magnitude of 2 means that the line will be less steep than the previous line. We have to be careful with the y-intercept, though. The general equation of y = mx + b specifically has a plus sign to the left of the b, but our example of y = -2x – 3 has a minus sign. Recall that subtracting is really the same as adding a negative. This means that we can rewrite our example as y = -2x + (-3), or at least think of it as such. We can now see that our y-intercept is actually -3.

136

We now know that one point on the line will be (0, -3). The slope of -2 means that the line will slope down and to the right, and the ratio of vertical change to horizontal change is 2:1. Stated another way, to get from one point on the line to the next, we go down 2, and right 1.

A typical exercise provides us with an equation of a line in slope-intercept form, and asks us to determine the value of y for a given value of x. Let's look at an example: For the equation $y = (\frac{1}{2})x + 4$, find the value of y when $x = 6$. All we do is substitute the given value of x in the equation and evaluate. We have $(\frac{1}{2})(6) + 4$ which is 7. This means that on the line represented by $y = (\frac{1}{2})x + 4$, if we go to the place where x equals 6, y will equal 7. Stated another way, we have learned that the point (6, 7) is a point on the line. We say that the point (6, 7) satisfies the equation of the line, since y equals 7 when x equals 6.

For practice, try graphing the line $y = (\frac{1}{2})x + 4$ using the method used earlier. We know that the y-intercept is 4 which means we can use (0, 4) as our starting point. The positive slope means the line will slant up and to the right, and the slope value of $\frac{1}{2}$ means that the ratio of vertical change to horizontal change is 1:2. This is not at all the same as 2:1. To get from one point on the line to the next, we will go up 1, and right 2. This will result in a line that is less steep than a line of slope 1 which makes a 45° angle like the one we saw earlier. If you plot the line in this manner, you will see that (6, 7) will be a point on it.

CONVERTING FROM THE STANDARD EQUATION FORM OF A LINE ($Ax + By = C$) TO ($y = mx + b$) FORM

A common task in this topic is to take what is called the **standard equation form** of a line ($Ax + By = C$), and algebraically convert it into the slope-intercept form ($y = mx + b$) that we've been working with. A demonstrative example will make this clearer.

The equation $6x + 2y = 8$ represents a line, but it doesn't look like the linear equations we've been working with. One concern is that we can't readily determine the slope and y-intercept of the line just by looking at the equation since it isn't in the slope-intercept format of $y = mx + b$.

Let's use our basic algebra techniques to "manipulate" the above equation so that it is in slope-intercept format. There is more than one correct way to do this, but we will work with the simplest. First, we don't want to have both the x and y on the same side since they are on opposite sides in the slope-intercept form of a line. We can "move" the 6x over to the right side by subtracting 6x from each side, thereby "cancelling" it on the left.

We now have $2y = 8 - 6x$. Notice that the 6x now has a minus sign since we subtracted 6x from both sides to "cancel" it on the left, and "move it over" to the right.

Next, observe that in the formula $y = mx + b$, the x term appears to the left of the b which is our y-intercept constant. We need to swap the positions of $8 - 6x$, but we must be careful, and many students get confused with the details. Remember the concept that a minus sign to the left of a value can be treated as a negative, and we can insert an implied addition operation sign once we've done so. The $8 - 6x$ becomes the equivalent $8 + (-6x)$. We can then use the commutative property of addition to swap the order to get $-6x + 8$.

We now have $2y = -6x + 8$. We need to "get rid of" the 2 to the left of the y since we need y by itself on the left. We can accomplish that by dividing both sides by 2, remembering that the 2 will distribute over both terms on the right. We end up with $y = -3x + 4$, from which we can see that the slope is -3 and the y-intercept is 4.

Admittedly, this whole process can be confusing, and takes practice. It is no exaggeration to say that in order to do problems like this, you must be completely confident in handling basic algebra and basic arithmetic operations with signed numbers. If you aren't, go back and review those concepts until they are fully mastered.

For more practice, let's convert $5x - 3y = -6$ to slope-intercept format. We want our x term to be on the right and not on the left with the y term, so let's subtract 5x from each side. We now have $-3y = -6 - 5x$. The minus sign to the left of the 3y has effectively become a negative sign. Swap the order of the terms on the right side to get $-5x - 6$, remembering that they really have an implied addition sign between them.

We now have $-3y = -5x - 6$. Divide both sides by -3 to get y by itself on the left, resulting in $y = (5/3)x + 2$. From this, we can see that the slope is (5/3) and the y-intercept is 2.

Questions involving this task are common because they incorporate many important concepts all at once. There is no way to do these problems accurately unless you fully understand all of those concepts. Note that this topic plays a large role in later geometry, so if you are moving ahead in math, now is the time to master this topic.

IMPORTANT POINTS TO REMEMBER

Many easy test questions are based on plotting or identifying points on a coordinate plane. Carefully review the related section until the matter is completely clear to you. Make sure that there is no point on a graph that you would have trouble identifying, and that there is no ordered pair that you would have trouble plotting.

All of the slope formula variants are just "change in y over change in x." To determine the slope between two points, pick either point to be

Point A, and the other point to be Point B. Maintain that consistency while you subtract the y values and the x values.

A vertical line has an undefined or no slope, and a horizontal line has a 0 slope. Those terms are not interchangeable even though they may sound similar.

The slope-intercept form of a line is represented by the formula $y = mx + b$ where m is the slope and b is the y-intercept. Thoroughly review the sections in this chapter on the significance of those terms.

The standard equation form of a line is represented by the equation $Ax + By = C$ where A, B, and C are constants. To determine the slope and y-intercept of the given line, we must first algebraically rearrange the equation into slope-intercept form.

For practice, try these exercises:

1) Find the slope of the line which goes through points of coordinates (2, -6) and (-6, -8)?
2) Find the slope of the line which goes through points of coordinates (-1, -12) and (-4, 6)?
3) What is the slope of a vertical line?
4) What is the slope of a horizontal line?
5) What is the y-intercept of the line represented by $y = 5x - 6$?
6) What is the slope of the line represented by the equation $y = x - \frac{3}{4}$?
7) Convert to slope-intercept format: $27x + 9y = 18$
8) Convert to slope-intercept format: $6x - 5y = -15$

SO NOW WHAT?

Start at the beginning, refocusing on any concepts that you weren't fully comfortable with. Allow at least a day to go by without studying, then take the end-of-book test without referring back to the book.

End-of-Book Test

Take this self-test after you have read the material in this book, but not immediately after. Use it to determine how much of the material you are retaining, and what concepts you haven't yet fully internalized. Don't be concerned about how many questions you get right or wrong. Just make sure you understand why the right answers are right and the wrong answers are wrong.

After completing this test and checking your answers, review the concepts in this book with an emphasis on those that you either forgot or had trouble with. If you have questions or need help, contact me via my website.

1) What are four ways of representing "three times y?"
2) Evaluate $-3x - 11$ when x is -2
3) Combine like terms: $-7x^2 - x - 9 + 8x^3 - 3x + 5x^2$
4) Combine like terms: $5x^3 - 5x^2 + 5x - 5$
5) Simplify: $a^4 + a^4$
6) Solve for x: $x + 5 = -8$
7) Solve for x: $x - 2 = -7$
8) Solve for x: $-3x = 12$
9) Solve for x: $18x = 9$
10) Solve for x: $\frac{x}{-2} = -8$
11) Solve for x: $\frac{2}{5}x = 40$
12) Solve for x: $\frac{1}{8} + x = \frac{3}{4}$
13) Solve for x: $x - 2.93 = 8.6$
14) Solve for x: $-7 - (-9) + 8 = 5x - 8x + 13x$
15) Solve for x: $-5x - 6 = 4$
16) Solve for x: $-8x - 5 = -10x + 3$

17) A student received grades of 72, 86, 91, 62, and 53. What grade must s/he get on the next test to have a mean of 77?

18) A person invests $2700 for 9 months in an account that offers an APY of 1.3%. How much money in total will s/he have at the end of the time period?

19) A person wants to earn $500 on a $3200 investment. If s/he is willing to invest the money for 6 years, what is the minimum interest rate s/he must find?

20) Solve for x: $\frac{2}{9} = \frac{7}{x}$

21) Solve for x: $4(x + 7) = 22$

22) Solve for x: $-9(x - 3) = -18$

23) Simplify: $3x^2 \cdot 7x^4$

24) Simplify: $x^3 + x^8$

25) Simplify: $9x^7 \div 27x^4$

26) Simplify (with positive exponent): $21x^5 \div 7x^7$

27) Simplify: $(x^5)^7$

28) Evaluate: 2^0

29) Evaluate: 5^1

30) Simplify: $8xy^2z^3 \cdot 7x^2y^4z^8$

31) Distribute and simplify: $7x^2(4 + 6x^5)$

32) Distribute and simplify: $-5x^3(2x - 9)$

33) Solve for x: $x - 5 \le -8$

34) Solve for x: $-21x > 14$

35) Solve for x: $5x \ge -10$

36) Multiply with FOIL: $(x - 7)(x + 3)$

37) Factor: $x^2 - 12x - 13$

38) Factor: $x^2 - 144$

39) Factor: $24x^7 - 36x^{11}$

40) Factor and simplify: $x^2 - 12x + 36$

41) State the degree measurement/range of these angles:
a) right; b) straight; c) acute angle; d) obtuse

42) Two angles are complementary. One measures 26.7 degrees. What is the degree measure of the other?

43) Two supplementary angles are represented by the expressions $(2x - 15)$ and $(6x + 43)$. Find the degree measure of the smaller one.

44) Find the area of a 7 mm by 8 mm rectangle.

45) Find the area of a square of side 16 yards.

46) Find the area of a right triangle with legs of 3 and 4 inches, and a hypotenuse of 5 inches.

47) Find the area of a parallelogram with a base of 7 cm, slanted sides of 4 cm each, and a height of 3 cm.

48) Find the area of a trapezoid whose illustration (not drawn to scale) depicts it as having bases of 8 ft. and 9 ft., height 2 ft., and slanted sides of 3 ft. and 5 ft.

49) Find the perimeter of the above trapezoid.

50) True/False: $\pi = 3.14$

51) True/False: $\pi = 22/7$

52) Find the area of a circle of diameter 9 m.

53) Find the circumference of a circle of radius 3.7 in.

54) Find the volume of a rectangular solid of dimensions 2 by 7.2 by 3.5 feet.

55) Find the volume of a cube of edge 11.1 mm.

56) Find the volume of a cylinder with circular base of area 15 cm^2 and height of 7 cm.

57) Find the volume of a triangular prism of depth 12 in. whose triangular base has a height of 3 in. and a base of 5 in.

58) A circle's radius is quadrupled. How many times larger is the area of the resulting circle?

59) A rectangular solid has both its length and depth doubled, and its width tripled. How many times larger is the area of the resulting solid?

60) A triangle has angles of 27.3 and 18.6 degrees. What is the degree measurement of the third angle?

61) The angles of a triangle are represented by the expressions $(7x - 12)$, $(4x + 4)$, and $(3x - 8)$. Find the degree measure of the largest one.

62) How many degrees does each angle measure in an equilateral triangle?

63) An isosceles triangle has base angles represented by the expressions $(7x + 5)$ and $(6x + 11)$. Solve for x.

64) A 10 ft. tall pole casts a 7 ft. shadow. How tall is a nearby tree if its shadow measures 10.5 ft at the same time of day?

65) A right triangle has legs of 9 and 12 cm. Find the hypotenuse w/o using the Pythagorean Theorem.

66) A right triangle has a leg of 7 ft., and a hypotenuse of 11 ft. What is the length of the other leg rounded to the nearest tenth of a foot?

67) Find the slope of the line which goes through points of coordinates (3, -7) and (-5, -9)?

68) What is the slope of a vertical line?

69) What is the slope of a horizontal line?

70) What is the y-intercept of the line represented by the equation $y = 3x - 5$?

71) Find the slope of the line represented by the equation $y = -x + \frac{1}{2}$

72) Convert to slope-intercept format: $7x + 5y = 10$

PREPARING FOR A STANDARDIZED EXAM

The first two books in the series offer many practical tips on how to prepare for math exams, but some additional points are worth mentioning. When studying for a standardized exam, it is ineffective to obsess over the few sample questions typically included in the official exam booklet or on the website of the exam company. While it is certainly beneficial to get a sense of the exam format, and how you will be expected to indicate your answers, those are just logistical issues.

The best way to prepare for an exam is to ensure that you have a solid mastery of the concepts you will be tested on. Even if the questions on your actual exam turn out to be very similar to any sample questions provided, just slight alterations are enough to test if you truly understand the underlying concepts, and are not just attempting to solve the questions by rote. Maintain this mindset while studying old exams. Treat them as extra practice, but not as an advance copy of your exam.

While it is easy to determine in advance what general topics are covered on a given standardized exam, do not assume that familiarity with a topic is the same as having truly mastered it. For example, a student may note that his/her scheduled exam covers the topic of "integers," and respond with, "Yeah, I studied integers, I know them." If nothing else, try to get more specific information about what exactly will be tested, but in any case, just spend as much time as possible studying and reviewing any and all related material.

A math exam is not the same as a trivia contest in which breadth of knowledge may be advantageous. On a math exam, either you know how to do a question or you don't. You don't get any points for recognizing a topic or for having a vague idea of how to proceed with a question. For most basic standardized exams, you are better off studying concepts of basic math than rapidly skimming through a great deal of more advanced material.

If you have truly mastered all of the material in the first book and in this one, you have a very solid foundation for most standardized tests, and will likely achieve your goal. Even if your specific exam includes a few advanced topics that aren't covered until later in the series, don't obsess over those topics unless you truly have time to learn them. You are usually better off being 100% prepared for 90% of the topics than the other way around. ▴

Answers to Exercises and Self-Tests

Please read the section in the Introduction on typos and errors. Remember that you can contact me with any questions, and visit my website for help and information. On request, I can provide additional practice exercises for any topic, although you should also try making up your own practice exercises. Make sure that you understand why your right answers are right, and why your wrong answers are wrong.

ASSESSMENT TEST OF PREREQUISITE MATERIAL

1) +, ×
2) 63
3) -2
4) 8, 16, 24, 32, 40, 48, 56, 64, 72, 80
5) 13 ⅗
6) <
7) -20
8) 17,201,018,040
9) 38,700
10) 250,000
11) ±2
12) 706.94
13) >
14) >
15) 37/1000
16) 0.43
17) 0.12
18) $0.\overline{6}$
19) True
20) True
21) <
22) >
23) 61,800
24) 0.0234
25) 23.457
26) 60.00
27) 13/6
28) 53/4
29) It is 1/1000 the size
30) 48 cents
31) 744
32) 45
33) 117/2
34) 213

35) -6/1
36) 12
37) 12
38) 23/37
39) A whole number: pos., neg., or zero
40) -4
41) 7
42) -13
43) -8
44) -5
45) -17
46) -4
47) -6
48) -5
49) -54
50) 56
51) -5
52) 8

53) 256
54) Undefined
55) 625
56) 1, 2, 29, 58
57) 1, 29
58) 0
59) Undefined
60) 9
61) 21/11
62) 7/12
63) False
64) False
65) False
66) True
67) 9/19
68) 3/16
69) 49/81
70) 1

CHAPTER ONE

1) $7z, (7)(z), 7 \cdot z, 7 \times z$
2) -1
3) $4x^3 - x^2 - 3x - 2$
4) $7x^3 + 5x - 7x + 5$
5) $3a^3$

CHAPTER TWO

1) -20
2) 16
3) -5
4) ½

5) 9
6) 100
7) 9/16
8) 11.73

CHAPTER THREE

1) 3/10
2) -12
3) 2
4) 86
5) $3645

6) 3.9% (rounded)
7) $47.\overline{6}$
8) 2.4
9) 5.5

CHAPTER FOUR

1) $32x^8$
2) $x^4 + x^9$
3) $\frac{1}{2}x^6$
4) $4/x^4$
5) x^{48}

6) 1
7) -6
8) $54x^4y^8z^{13}$
9) $40x^3 + 56x^9$
10) $-12x^5 + 44x^4$

CHAPTER FIVE

1) $x \le -7$
2) $x < -\,^2/_3$
3) $x \ge -2$

4)

5)

CHAPTER SIX

1) $x^2 - 4x - 12$
2) $x^2 - 13x + 40$
3) $(x + 2)(x + 3)$
4) $(x - 4)(x - 3)$
5) $(x - 8)(x + 8)$

6) $(x - 15)(x + 15)$
7) $5x^8(2 - 3x^4)$
8) $4x^6(1 + 2x^{11})$
9) $(x + 5)^2$
10) $(x - 7)^2$

CHAPTER SEVEN

1) 90; 180; btw. 0 and 90; btw. 90 and 180
2) 51

3) 163
4) -2

CHAPTER EIGHT

1) $63\ cm^2$
2) $625\ yd^2$
3) $30\ in^2$
4) $32\ cm^2$
5) $28.5\ ft^2$
6) False
7) True
8) $64\pi\ cm^2$
9) 10π ft.
10) $121.26\ ft^3$
11) $121.26\ ft^3$
12) $178.5\ cm^3$
13) $240\ in^3$
14) 9
15) 36

CHAPTER NINE

1) 39.4
2) 19
3) Two equal angles, two equal sides
4) ⅓
5) 9.3 ft. (rounded)
6) 26
7) $\sqrt{105} \approx 10.2$

CHAPTER TEN

1) 1/4
2) -6
3) Undefined
4) 0
5) -6
6) 1
7) $y = -3x + 2$
8) $y = (6/5)x + 3$

END-OF-BOOK TEST

1) $3y$, $(3)(y)$, $3 \cdot y$, $3 \times y$
2) -5
3) $8x^3 - 2x^2 - 4x - 9$
4) $5x^3 - 5x^2 + 5x - 5$
5) $2a^4$
6) -13
7) -5
8) -4
9) ½
10) 16
11) 100
12) 5/8
13) 11.53

14) 1

15) -2

16) 4

17) 98

18) $2726.33

19) 2.6% (rounded)

20) 31.5

21) -1.5

22) 5

23) $21x^6$

24) $x^3 + x^8$

25) $(1/3) x^3$

26) $3/x^2$

27) x^{35}

28) 1

29) 5

30) $56x^3y^6z^{11}$

31) $28x^2 + 42x^7$

32) $-10x^4 + 45x^3$

33) $x \le -3$

34) $x < -2/3$

35) $x \ge -2$

36) $x^2 - 4x - 21$

37) $(x - 13)(x + 1)$

38) $(x - 12)(x + 12)$

39) $12x^7(2 - 3x^4)$

40) $(x - 6)^2$

41) 90; 180; btw. 0 and 90; btw. 90 and 180

42) 63.3

43) 23

44) $56 \, mm^2$

45) $256 \, yd^2$

46) $6 \, in^2$

47) $21 \, cm^2$

48) $17 \, ft^2$

49) $25 \, ft$

50) False

51) False

52) $20.25\pi \, in^2$

53) $7.4\pi \, in$

54) $50.4 \, ft^3$

55) $1368 \, mm^3$ (rounded)

56) $105 \, cm^3$

57) $90 \, in^3$

58) 16

59) 12

60) 134.1

61) 86

62) 60

63) 6

64) 15 ft.

65) 15 cm.

66) $\sqrt{72} \approx 6.5$ ft.

67) $\frac{1}{4}$

68) Undefined

69) 0

70) -5

71) -1

72) $y = (-7/5)x + 2$

HOW TO GET MORE HELP

I maintain a free, compressive website at www.MathWithLarry.com including the means for students to e-mail me their math questions. That will continue with the publication of the *Math a Bit Easier* book series, although I'm working to redesign the site to better align it with the books. New content will be added as students ask questions about the material. My goal is for the website to serve as an interactive book companion so that students' questions can be addressed. My website and question-answering service will always be free for all.

You can do it! You can achieve the math goals you have set for yourself, but doing so will certainly take time and effort. Much of this will be in the form of reviewing and filling in gaps in earlier math material that may not have been fully mastered at the time. Contact me via my website if you have questions about the material, or would like to discuss your academic situation. Study hard and believe in yourself! ☺

ABOUT THE AUTHOR

Larry Zafran was born and raised in Queens, NY where he tutored and taught math in public and private schools. He has a Bachelors Degree in Computer Science from Queens College where he graduated with highest honors, and has earned most of the credits toward a Masters Degree in Secondary Math Education.

He is a dedicated student of the piano, and the leader of a large and active group of board game players which focuses on abstract strategy games from Europe. He lives with his fiancée in Cary, NC where he works as an independent math tutor, author, and webmaster.

www.LarryZafran.com

22783016R00082

Made in the USA
Lexington, KY
13 May 2013